REPORT

Launching the Qatar National Research Fund

Shelly Culbertson • Michael G. Mattock • Bruce R. Nardulli • Abdulrazaq Al-Kuwari
Gary Cecchine • Margaret C. Harrell • John A. Friel • Richard E. Darilek

Sponsored by the Qatar Foundation

RAND RAND-QATAR POLICY INSTITUTE

This research was sponsored by the Qatar Foundation for Education, Science, and Community Development and was conducted under the auspices of the RAND-Qatar Policy Institute and the Transportation, Space, and Technology Program within RAND Infrastructure, Safety, and Environment.

Library of Congress Cataloging-in-Publication Data is available for this publication.

ISBN: 978-0-8330-5341-4

The RAND Corporation is a nonprofit institution that helps improve policy and decisionmaking through research and analysis. RAND's publications do not necessarily reflect the opinions of its research clients and sponsors.

Published 2012 by the RAND Corporation
1776 Main Street, P.O. Box 2138, Santa Monica, CA 90407-2138
1200 South Hayes Street, Arlington, VA 22202-5050
4570 Fifth Avenue, Suite 600, Pittsburgh, PA 15213-2665
RAND URL: http://www.rand.org/
To order RAND documents or to obtain additional information, contact
Distribution Services: Telephone: (310) 451-7002;
Fax: (310) 451-6915; Email: order@rand.org

Preface

In 2004, at the request of the Qatar Foundation for Education, Science, and Community Development (QF), the RAND-Qatar Policy Institute (RQPI) developed a proposal for the design of the Qatar National Research Fund (QNRF) and drafted accompanying business and implementation plans. The QF Board of Directors approved the design and plans,[1] and then, from 2006 to 2008, QF asked RQPI to assist in further developing and carrying out plans for the start-up of QNRF. Since then, QNRF has grown into a research funding organization that has allocated about $500 million for research in Qatar, in partnership with other countries.

QNRF—the first research-funding organization of its kind in the Middle East—has recently passed its five-year anniversary, and this report takes that occasion to present an overview of its launch, including the design and implementation of its first programs, from August 2006 through January 2008. The report describes the original analysis behind the programs, policies, planning methods, and decisions and discuses QNRF's experience with the first grant cycles, early results from the programs, and initial improvements upon them. QNRF has developed in many areas since that time, but the report should be a resource for individuals in Qatar who would like to know more about the planning process behind starting QNRF; policy leaders in other nations (particularly in the Middle East) who are interested in starting a research-granting organization; researchers seeking funding from QNRF; analysts and consultants who may be asked to tackle similar tasks; and persons interested in science and technology policy and educational and research infrastructure in emerging markets.

This report will be of interest to officials of QF, QNRF, and the government of Qatar who are involved in making decisions on research issues related to the country's overarching vision for its future. It should also interest the broader research community in Qatar and elsewhere that has followed QNRF's development to date.

This research was conducted under the auspices of RQPI and the Transportation, Space, and Technology Program (TST) within RAND Infrastructure, Safety, and Environment (ISE).

The RAND-Qatar Policy Institute

To study some of the most important issues facing the Middle East, RAND and the Qatar Foundation for Education, Science and Community Development formed a partnership that

[1] The business plan was drafted by a team headed by Debra Knopman, a Vice-President of the RAND Corporation and Director of RAND Infrastructure, Safety, and Environment; the implementation plan was drafted by a team headed by D. J. Peterson, a senior RAND researcher.

in 2003 established RQPI in Doha, Qatar. RQPI is an integral part of Education City, which is being developed by QF under the leadership of Her Highness Sheikha Moza Bint Nasser. Education City is a community of institutions—both K–12 and universities—contributing to education and research in Qatar and the Gulf region. RQPI is a regional office that facilitates delivery of the full range of RAND's capabilities to clients in North Africa, the Middle East, and South Asia—roughly, from Mauritania to Bangladesh.

Further information

For further information about this report, other RQPI work on QNRF, or RQPI, contact:

Dr. Obaid Younossi, Director
The RAND-Qatar Policy Institute
P.O. Box 23644
Doha, Qatar
Tel: 00974-4454-2500/02
Email: obaid@rand.org

Foreword

The story of QNRF dates back to early 2003, when Her Highness Sheikha Moza Bint Nasser, Chairperson of the Qatar Foundation for Education, Science and Community Development (QF), initiated the idea of a research fund, leading to the founding of QNRF.

As part of this initiative, QF tasked the RAND-Qatar Policy Institute (RQPI) with conducting a study into the formation of QNRF over the period from February 2004 through June 2004, and in August 2005, the QF Executive Board of Directors approved the business and implementation plans that RQPI had developed as part of its study.

One year later, in August 2006, QNRF's start-up team arrived in Doha. Under the guidance of Dr. Amir Al-Saadi, Research Advisor to Her Highness Sheikha Moza Bint Nasser and QNRF Project Supervisor, the team took the first steps of a long and very challenging journey. That journey included the essential tasks of developing the funding programs, building the infrastructure, recruiting staff, putting in place key policies and procedures, and coordinating efforts with stakeholders to accomplish the goal of making QNRF a reality.

At that time, the start-up team had two approaches available to achieve its goal. The first approach was a conventional step-by-step process, starting with recruiting and training staff and developing procedures and infrastructure prior to launching the research-funding programs. This process would have taken a few years to accomplish. The alternative was to take a sink-or-swim approach and immediately jump in at the deep end, launching at least some of the funding programs on a fast-track basis. "Sink" was not going to be an option.

This volume describes the first year and a half (August 2006 to February 2008) of the journey, when we at QNRF consolidated our efforts with our colleagues at RAND and kick-started the process.

Needless to say, QNRF has come of age since then, growing into a professional, globally recognized funding agency with clear, well-established procedures and guidelines, as well as state-of-the-art online application, grant management, and peer-reviewing solutions. By early 2012, QNRF had implemented 11 cycles of the Undergraduate Research Experience Program (UREP), five cycles of the National Priorities Research Program (NPRP), two cycles of the Young Scientists Research Experience Program (YSREP), and two cycles of the Senior School Research Experience Program (SSREP).

QNRF recently passed its five-year anniversary, and our colleagues at RAND have documented or, rather, archived the early stages of QNRF's adolescence to deliver the message to other countries that when the will and the commitment are there, nothing can stand in the way of achieving one's goals.

Dr. Abdul Sattar Al-Taie, QNRF Executive Director
Doha, Qatar
February 2012

Contents

Figures and Tables

Figures

Tables

Summary

In its first five years of operation, the Qatar National Research Fund (QNRF) grew from a small start-up to a research-funding institution that had awarded about $500 million in research grants. It was the first institution of its kind in the Middle East, starting in 2006 with a small staff and a broadly outlined vision. Within months of starting, it had launched its first program, the Undergraduate Research Experience Program (UREP). In spring 2007, it launched the National Priorities Research Program (NPRP), its primary funding vehicle.

In its first 11 funding cycles, the UREP—in which undergraduates enrolled in Qatar's universities participate in research projects mentored by faculty—made awards totaling about $15 million to about 1,500 students in all of Qatar's universities. In its first five rounds of funding, the NPRP, QNRF's grant program for professional researchers, awarded about $485 million to research teams in Qatar that partnered with researchers from institutions in more than 30 other countries. Through these programs, QNRF has also laid the foundation for a domestic research infrastructure in Qatar to support the growing research communities in Qatar University and Doha's Education City, home to six branch campuses of U.S. universities.

While QNRF has developed significantly since that time, this report discusses its start-up from August 2006 through January 2008, including the initial analysis, decisions made, implementation, and early results. During this period, the RAND-Qatar Policy Institute (RQPI) served as advisor and worked hand-in-hand with QNRF's director and growing staff to provide analysis, aid in project planning, design programs, contribute to making pivotal policy decisions, draft key documents, make programs operational, and ensure quality in processes and products. Figure S.1 shows the initial time line of the start-up of QNRF.

In 2004, before the start-up, at the request of the Qatar Foundation for Education, Science, and Community Development (QF), RQPI had created an initial design for QNRF as well as business and implementation plans. The new organization was envisioned to become an internationally recognized institution that would use research as a catalyst for "expanding and diversifying the country's economy; enhancing the education of its citizens and the training of its workforce; and fostering improvements in the health, well-being, environment and security of its own people and those of the region" (Greenfield et al., 2008).

The proposed institution had three goals:

- Building human capital
- Funding research in the interest of Qatar, the region, and the world
- Raising Qatar's profile in the international research community.

The first goal, building human capital, was the most important in the early phases.

Figure S.1
Initial Start-Up Time Line of QNRF

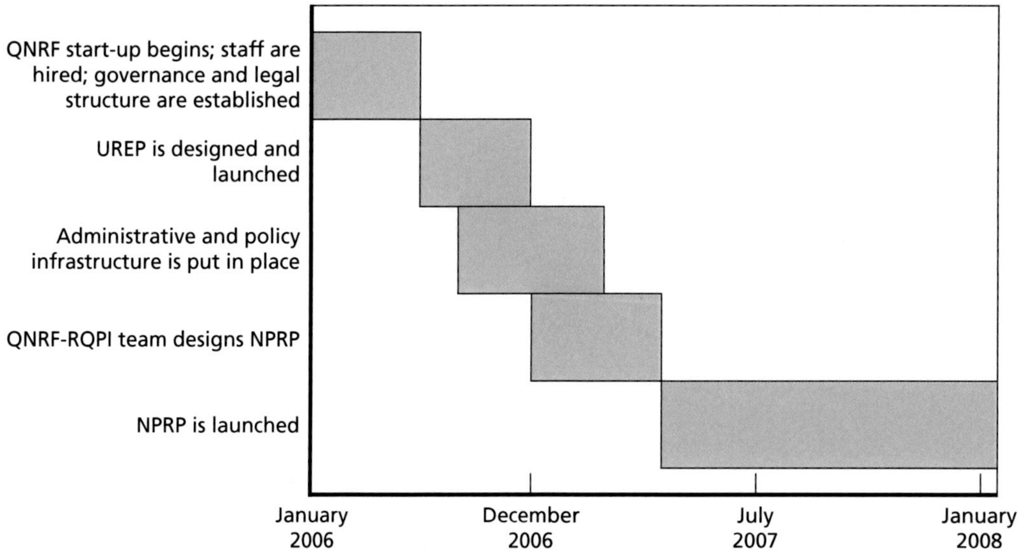

RAND *TR722-S.1*

Laying the Foundation for Ongoing Operations: Creating an Infrastructure for Governance and Management

Before QNRF could begin operations and launch its first grant programs, several basic building blocks had to be in place. The fund's legal standing needed to be determined, and an arrangement for governance had to be created. An organizational structure also needed to be worked out, and staff had to be hired.

Choosing QNRF's Legal Form

The QF Board of Directors opted to make QNRF a subsidiary of its parent, QF. An alternative option had been to make it an independent legal entity with its own governing board and guidance from QF, but the QF leadership decided that the new research fund would be better served with the leadership and financial support available to it as a QF center. It was also decided that once QNRF had sufficient staffing, policies, and experience as an organization, its governance would transition to a board-led model, still under QF authority. Following this plan, an interim steering committee was appointed in lieu of a board, with members affiliated mainly with QF and institutions in Education City. The steering committee could meet more regularly than a board, could make decisions more quickly, and would allow greater flexibility during the start-up period.

Designing an Organizational Structure and Hiring Staff

A core team of QNRF staff was hired in late 2006. This team consisted of a start-up director, a special-projects officer, and two program managers. Given QNRF's ambitions, however, a detailed organizational structure and additional staff were needed in short order. The QNRF-RQPI team decided to use the model of a matrix organization, in which employees would assume different roles, depending on the organization's needs at any given time. This allowed a small number of employees to take on an array of responsibilities as circumstances required.

Lessons Learned from QNRF's Governance and Management Structure

The QNRF-RQPI team took away two key lessons from the experience of building QNRF's structure: First, QNRF remained understaffed during the start-up because the organizational structure was too "lean" and because of difficulties in hiring qualified staff. Second, the flexibility of the steering committee was very important during the start-up, so QF decided to keep the steering committee, although it planned to add a higher-level governing board at some point in the future.

Guiding Principles for the Design of QNRF's Research Programs

Having laid the foundation for operations, the QNRF-RQPI team turned to designing QNRF's first programs. As a first step, it worked out six guiding principles to which it could refer as it developed programs and policies:

- QNRF programs should aim to create a research culture in Qatar, focusing on building human capital.
- Program designs should include attractive incentives for researchers and institutions.
- Programs need one set of policies that can accommodate research in different parts of the world.
- Programs require "buy-in" (support and feedback) from participating institutions in order to effectively meet those institutions' needs.
- QNRF should learn from its own experiences in designing programs and should make improvements.
- QNRF policies should be clear, transparent, and consistently applied.

Planning and Launching the Undergraduate Research Experience Program

In keeping with the guiding principle that QNRF should learn from its experiences as it designs successive grant programs, the QNRF-RQPI team decided to first launch the UREP, QNRF's funding vehicle for faculty-mentored undergraduate research projects. The team knew that the UREP would be significantly smaller than the NPRP. Tackling it first would allow the new organization room for experimentation and trial-and-error experience with a lower-stakes program, yet the process of designing its policies and administration—creating the program with university input, writing a request for proposals (RFP), soliciting applications, setting up a peer-review process and scoring system, and finding peer reviewers—would be a smaller-scale model for the NPRP. In addition, the UREP would be manageable for QNRF's small start-up staff.

Because there was a push to demonstrate QNRF's viability as an organization as quickly as possible, the QNRF-RQPI team began designing the UREP in the first months of the start-up phase, while it was establishing the fund's basic operational structure. Recognizing that students form Qatar's future workforce, the QNRF-RQPI team wanted the UREP to create opportunities for Qatar's undergraduate population. Therefore, it designed a program that would award grants to faculty at universities in Qatar to direct research projects staffed by undergraduates. The mentoring involved would supplement normal classroom instruction, and "learning by doing" would improve the quality of participating students' education, give them practical collaborative experience, and perhaps inspire them to continue their studies at the graduate level.

The first UREP RFP was issued in October 2006, a few months after QNRF itself came into existence. Faculty at universities in Qatar submitted a total of 120 proposals—roughly four times the response rate anticipated in QNRF's original business plan. Peer reviewers recruited from institutions around the world rated more than half of these submissions "very good" or "excellent," and 61 proposals received funding. In the UREP's first round, QNRF awarded a total of $1,322,000 in grants to the universities that submitted proposals. The UREP had similar results in its second and third funding cycles. As QNRF's first program, the UREP had demonstrated that QNRF could start and run a research-granting program.

Planning the National Research Priorities Program

With policies in place and the experience gained in running the UREP, QNRF could turn its focus to its main vehicle for supporting research in Qatar, the NPRP. The scale envisioned for the NPRP called for the QNRF-RQPI team to do considerable planning and to make a number of key decisions. The most important planning priorities were

- The nature and structure of the program
- The program's research priorities
- The peer-review process
- The RFP
- Intellectual property (IP) policies.

The original concept for the NPRP in the 2004 business and implementation plans was a program that would fund no more than 16 large multi-investigator grants at universities in Qatar. However, the diversity and creativity of the proposals submitted for the UREP were encouraging, and the process of making award decisions on the basis of merit in open competition had had promising results. With that in mind, the QNRF-RQPI team decided that taking the same approach with the NPRP might better support creation of a dynamic research community in Qatar and achieve more of QNRF's objectives. The team thus decided that the NPRP would issue a general RFP and evaluate the proposals on merit, without considering the institutions involved.

A similar rethinking took place with regard to the research topics the NPRP would fund. It had originally been thought that QNRF would award NPRP grants to projects on specific, preselected research topics that are pivotal to Qatar's national interests, but when the time came to choose these research priorities, it proved inappropriate for QNRF to decide what they should be. Choosing priorities meant determining what would not be funded as well as what would be funded. Deciding among these kinds of trade-offs would be politically sensitive; would take research, time, and negotiation among stakeholders to produce a definitive list; and would require endorsement by higher levels of authority in Qatar. Thus, prioritizing specific and exclusive research topics could potentially delay the launch of the NPRP indefinitely.

Consequently, the team decided to view NPRP research priorities through the lens of building capabilities rather than funding specific areas. The NPRP would accept proposals of applicants' choosing, in any research field. Award decisions would then be based on the degree to which the proposed projects would help meet QNRF's goals of building human capital; funding research in the interest of Qatar, the region, or the world; and raising Qatar's profile in the international research community. Building human capital through creating research capability in Qatar was the first priority. In later cycles, the program could broaden its focus to achieve progress on the other goals.

To promote building human capital in Qatar, the team integrated a carefully chosen set of incentives and requirements into the RFP:

- It encouraged intellectual freedom by allowing researchers to submit proposals for topics of their own design and determined technical merit through competition.
- It provided a non-exclusive list of suggested sample topics to motivate researchers' thinking about research areas of importance to Qatar.
- The majority of the budget spent and at least half of the work done would be in Qatar, and certain key personnel for projects would have to reside in the country.
- It allowed about one-third of the project budgets to be administered by research institutions abroad, as research collaboration is a very effective way to build human capital.
- It encouraged collaboration among institutions in Qatar.
- It provided incentives for participating institutions to establish needed policies and infrastructure to support research in Qatar.
- Applicants had to submit a letter of intent before proposals were due (the QNRF-RQPI team envisioned this as a way to get a head start on the task of lining up peer reviewers).

To facilitate the peer-review process (which would rely on researchers in respected research organizations around the world and not in institutions in Qatar that were eligible for the grants) and to help in decisionmaking in awarding NPRP grants, the QNRF-RQPI team instituted a "bin" system for the NPRP. When an applicant submitted a proposal, he or she was asked to choose one of seven bins (based on research discipline—for example, industry and engineering or social sciences) in which the proposal would compete. The bin system offered a number of advantages: Proposals in the same very general domain would compete only with each other, enabling QNRF to ensure diversity in the topics funded, and grants would go to the best proposals in each discipline. This would make it easier for QNRF staff to select peer reviewers. It also gave QNRF a means of prioritizing different disciplines or bins for funding if at some point in the future it should choose to do so.

Developing an IP policy was essential for the NPRP. The number and variety of stakeholders that might be involved in a QNRF grant—individual researchers, branch campuses in Education City, Qatar University, universities in the United States and Europe, private companies, and QF—made for a very complicated IP environment. Different stakeholders had varying IP policies and interests.

RQPI recommended an IP solution serving several purposes:

- To create an environment that would encourage innovation

- To permit compatibility between QNRF policies and the policies of QNRF grantees' home institutions
- To incentivize institutions to establish IP infrastructure
- To support the goal of generating revenue from IP.

The solution specified that when a grantee's home institution has an IP policy, QNRF should adopt a compatible policy. For institutions that do not have a policy in place, QNRF should utilize the prevailing international model, used throughout the United States and Europe. In this model, the researcher's home institution owns the rights to the IP, while the revenue is split evenly among the researcher, the researcher's department, and the researcher's institution (with a small portion of the last third going to QF). QNRF accepted this as a guideline, then negotiated individual agreements with grant recipients.

Launching the National Research Priorities Program

With the initial NPRP design complete, QNRF released the program's first RFP in April 2007. The QNRF-RQPI team conducted extensive outreach and heavily publicized the program in an effort to ensure the success of the first funding cycle. Taking a calculated risk, QNRF developed a basic online application system. More than 200 proposals were submitted by the deadline of August 2007, and after administrative review, 175 proposals moved on to peer review.

QNRF set a target of obtaining five peer reviewers per proposal, a considerable challenge, since QNRF had compressed the time line for making award decisions, wanting to make announcements by December, and the number of QNRF staff was limited. In addition, QNRF had set the bar for peer-reviewer qualifications very high, with requirements more stringent than those at many academic journals. The QNRF-RQPI team divided the work of finding enough qualified peer reviewers and also enlisted help from research-granting organizations that already had peer-reviewer databases.

By mid-November, the QNRF-RQPI team had lined up reviewers for most of the proposals. The effort went far toward meeting QNRF's original target: Most of the 175 proposals had four or more reviewers, while only 33 had three reviewers.

The QNRF-RQPI team recognized that it was essential in this first round of NPRP funding to set a precedent for transparency in the award process, as the impression created at the start could gain or lose the confidence of the research community. Consequently, the team decided to base funding solely on the numerical scores submitted by the peer reviewers, rather than using a panel process or having a committee make funding match certain research priorities. The team also opted to use an absolute, across-the-board standard score to make final decisions about which proposals to fund, instead of a relative standard in which proposals would be evaluated with respect to competing proposals in their bin. As it turned out, the first-round proposals had fairly even scores across all of the bins. Whether or not to determine different cut-off scores for each bin was not an issue.

The NPRP's first grants were awarded in December 2007. Of the 175 proposals that went to peer review, 47 were funded, for a total of $25 million over three years. At least one grant was awarded in each of the seven bins, and each major institution that submitted a proposal received a share of the funding.

Looking Ahead for QNRF

With QNRF, QF has taken solid steps toward achieving its goal of making Qatar a center of innovative education and research. In a short period of time, QNRF has grown from a few people to a grant institution with fully operational multimillion-dollar programs and the groundwork of a domestic research infrastructure in place. Since that time, QNRF has done much to expand and deepen its policies, build relationships with research institutions, and refine its programs.

Nevertheless, much remains to be done. Over time, QNRF will need to take next steps and meet a variety of fresh challenges. To truly become an internationally respected research-granting and managing organization, it will need to make changes in programs and policies to better meet the needs of Qatar's research community and goals.

Future developments will also require changes in QNRF's approach, management structure, policies, programs, and metrics. In 2007, the government of Qatar made a public commitment to significantly expand its investment in scientific research and technological development by dedicating 2.8 percent of its revenue to the effort. In 2012, QF began a stakeholder-driven process to develop a national research strategy for Qatar and focused research priorities. QNRF will be a key entity in supporting the new national research strategy. In light of that new strategy, QNRF will need to carefully examine its focus, governance and management infrastructure, policies, and programs, some already planned and some as yet unimagined. In addition, QNRF may benefit from developing further measurements of research outcomes. These are just a few of the many tasks still to be accomplished.

Acknowledgments

We would like to thank Her Highness Sheikha Moza Bint Nasser for providing the opportunity to help launch the Qatar National Research Fund. We would also like to thank Dr. Mohammed Fathy Saoud (President of QF), Dr. Abdulla Al-Thani (Vice President for Education at QF and President of Hamad bin Khalifa University), and Dr. Amer Al Saadi (QF Research Advisor and QNRF Project Supervisor) for directing and supporting RQPI's work on this project. We also give special thanks to QNRF Executive Director Dr. Abdul Sattar Al-Taie, QNRF Deputy Executive Director Dr. Nabeel Al-Salem, and the QNRF staff for their collaboration with RQPI researchers in the start-up effort. These staff include Senior Manager Mahmoud Talaat, NPRP Program Manager Noor Al-Merekhi, Program Manager for Biomedical Sciences and Health Dr. Thenaa Said, and IP and Innovation Specialist Dr. Imad Khadduri. Finally, we would like to thank the peer reviewers of this report, Dr. Bruce Don and Dr. Tora Bikson, for their valuable comments and insights.

Abbreviations

AES	Arab Expatriate Scientists
CFO	chief financial officer
CRDF	Civilian Research Development Fund
DARPA	Defense Advanced Research Projects Agency
DoD	Department of Defense
IP	intellectual property
IT	information technology
MOU	memorandum of understanding
NIH	National Institutes of Health
NPRP	National Priorities Research Program
NSF	National Science Foundation (United States)
PI	principal investigator
QF	Qatar Foundation for Education, Science and Community Development
QNRF	Qatar National Research Fund
R&D	research and development
RFP	request for proposals
RQPI	Rand-Qatar Policy Institute
UREP	Undergraduate Research Experience Program

CHAPTER ONE

Introduction

In its first five years of operation, the Qatar National Research Fund (QNRF) developed into a national institution for funding research conducted by organizations in Qatar in partnership with research organizations in more than 30 countries around the world. It is the first institution of its kind in the Middle East. To date (2012), it has awarded about $500 million in research grants through its two main programs, the National Priorities Research Program (NPRP) and the Undergraduate Research Experience Program (UREP). QNRF's grant programs support original, competitively selected research in a wide range of disciplines: the physical, life, and social sciences; engineering and technology; and the arts and humanities. It provides opportunities for researchers at all levels, from students to professionals, in the private, public, and academic sectors.

This report describes the design and start-up of QNRF, from July 2006 through January 2008. As a small start-up in late 2006, QNRF began with very little in terms of staff and structure. It had to develop rapidly into an organization that could manage multiple programs, numerous staff, hundreds of grant applications, and thousands of peer reviewers. At the beginning, the members of the planning team could all fit around one small conference-room table. By early 2008 (the end of the time frame for this report), a year and a half later, QNRF had administered three funding cycles of its earliest program, the UREP, and the first round of its largest granting program, the NPRP.

During the launch of QNRF from July 2006 to January 2008, the RAND-Qatar Policy Institute (RQPI) served as advisor and worked hand-in-hand with QNRF's director and growing staff to provide analysis, aid in project planning, design programs, contribute to making pivotal policy decisions, draft key documents, make programs operational, and ensure quality in processes and products.

This report describes the history and analysis behind the QNRF programs and policies as QNRF was establishing itself. It describes emerging results from the programs and some lessons learned. QNRF has developed significantly since its inception, and this report provides some broad recommendations for it, looking to the future.

The Origins of QNRF

The creation of QNRF was initiated by the leadership of the Qatar Foundation for Education, Science and Community Development (QF), an independent, nonprofit organization founded in 1995 by His Highness the Amir Sheikh Hamad bin Khalifa Al-Thani of Qatar. QF's mission is "to prepare the people of Qatar and the region to meet the challenges of an ever-changing world and to make Qatar a leader in innovative education and research" (Qatar Foundation, undated). QF has invested substantially in nationally oriented research and education. Recognizing the need for a national research fund to support its mission, in 2004 QF enlisted RQPI to help establish QNRF.

RQPI is a partnership of QF and the RAND Corporation.[1] It is an integral part of Education City, a growing community of educational institutions located in Qatar's capital, Doha.[2] Education City hosts branch campuses of several U.S. universities, including Carnegie Mellon University, Weill Cornell Medical College, Georgetown University School of Foreign Service, Texas A&M, Virginia Commonwealth University, and Northwestern University, as well as QF's own Faculty of Islamic Studies. Education City universities, along with the country's national university, Qatar University, form a hub of academic and research excellence for the nation.

At QF's request, RQPI developed a proposal for the design of QNRF and drafted accompanying business and implementation plans.[3] The RQPI team met with representatives of QF to develop a preliminary understanding of the foundation's original concept for QNRF's vision, mission, goals, and operating principles. Working with QF to refine this concept, the RQPI team first helped design a strategy for establishing and maintaining an infrastructure that would enable research to be conducted in Education City and elsewhere in Qatar (Greenfield et al., 2008).

QNRF was envisioned to become an internationally recognized institution that would use research as a catalyst for "expanding and diversifying the country's economy; enhancing the education of its citizens and the training of its workforce; and fostering improvements in the health, well-being, environment and security of its own people and those of the region" (Greenfield et al., 2008). To achieve this vision, QNRF had three goals:

- Building human capital
- Addressing national research needs
- Raising Qatar's profile in the international research community.

During this preliminary phase, the RQPI team undertook an analysis to further inform the emerging design. Seeking possible models for QNRF, it examined research-funding organizations and intellectual-property-rights regimes in the United States and other countries. It also consulted stakeholders in Qatar about the current status of research activities in the country, national needs for research, and potential opportunities and constraints.

The QF Board of Directors approved the plans that emerged from this work and later asked RQPI to provide advice on how best to implement them. RQPI worked closely with QF and the QNRF staff in an advisory role as plans for the official launch of the fund moved forward. In August 2006, QNRF's start-up phase began.

---

[1] Her Highness Sheikha Moza Bint Nasser of Qatar serves as co-chairperson of RQPI's Board of Directors, along with Michael Rich, RAND's President and Chief Executive Officer.

[2] QF is developing Education City under the leadership of Her Highness Sheikha Moza Bint Nasser.

[3] The business plan was drafted by a team headed by Debra Knopman, a Vice-President of the RAND Corporation and Director of RAND Infrastructure, Safety, and Environment; the implementation plan was drafted by a team headed by D. J. Peterson, a senior RAND researcher. The two plans are summarized in Greenfield et al. (2008).

Structure of This Report

Chapter Two describes the building of QNRF's governance and management infrastructure. Chapter Three discusses the guiding principles the QNRF-RQPI team used in designing the programs. Chapter Four addresses the planning and launching of the first program, the UREP. Chapter Five deals with the analysis and planning behind the NPRP. Chapter Six describes the results from the first NPRP cycle. Chapter Seven looks at next steps and future challenges for QNRF. Finally, the UREP request for proposals (RFP) is reproduced in Appendix A, the NPRP RFP is reproduced in Appendix B, and an overview of QNRF's strategy statements is provided in Appendix C.

CHAPTER TWO

Laying the Foundation for Ongoing Operations: Building QNRF's Governance and Management Infrastructure

To begin laying the foundation for QNRF, it was necessary to put together the basic elements of a working organization. After establishing the nature of the fund's legal standing and its relationship with its parent, QF, an arrangement for governance had to be set up. Finally, an organizational structure had to be created and staff hired. To support the preliminary decisions and processes, RQPI presented options to QF and the QNRF leadership and held discussions about advantages and disadvantages of various options; then QNRF and QF leadership made decisions and implemented them.

Choosing a Legal Form and Governance Arrangement

A pressing issue to be addressed immediately following the decision to create a research-funding institution in August 2006 concerned the kind of legal entity QNRF should be and the type of governance it should have. The original design had recommended a board-led governance model, similar to that of the National Science Foundation (NSF) in the United States and many other respected national and private foundations and programs (Greenfield et al., 2008; Buchanan, 2004; National Science Foundation, 2006). Because QNRF was in its infancy and was being launched by QF, it was important to carefully establish its legal relationship with its parent institution before proceeding.

Over a two-month period in September and October 2006, RQPI presented two options for QNRF's legal form for the QF Board of Directors to consider: QNRF could be either an independent legal entity with guidance from QF or a subsidiary of QF.

Option 1: An Independent Legal Entity with a Strong Governing Board

If QNRF was to be an independent legal entity, authority for policymaking, operational decisionmaking, and personnel and budget matters would be assigned to a QNRF governing board. QF would retain the authority only to appoint members of the QNRF governing board and establish the total level of QNRF funding. The Sidra Medical and Research Center, which was also founded by QF, uses a similar legal form and governance arrangement (Sidra, undated).

Option 2: A QF Center Fully Owned and Managed by QF

If QNRF was to be a subsidiary of QF, the QF Board of Directors could delegate certain responsibilities and authorities to a QNRF board but would retain ultimate authority. For example, it would have the power to change the responsibilities it delegates. A number of centers fully owned by QF, including the Qatar Science and Technology Park (QSTP) and Qatar

Academy, use this legal form (QSTP, undated; Qatar Academy, undated). Each of these centers has its own governing board, and the QF Board of Directors typically follows the recommendations of those boards. But final authority for contracting, appointing center directors, budgetary decisionmaking, and policymaking typically resides with the QF Board.

After extensive deliberations, QF concluded that QNRF would be better supported during its start-up as a QF center. It could utilize QF's existing legal and administrative infrastructure; reap the benefits of affiliation with Education City, such as name recognition and contacts with the American branch campuses and their home universities; and use QF's guidance, leadership infrastructure, and financial support during its formative years. It was decided at this time that governance would eventually transition to a board-led model when QF decided that QNRF had the appropriate staff and policies in place.

This decision obviated the immediate need for a separate QNRF governing board. Senior QF management decided that it would be better to establish an interim steering committee, which would offer a number of advantages over the more traditional board model during QNRF's start-up phase. A steering committee could be formed immediately and could be available to give counsel and direction on a more frequent basis (typically monthly) than a board could be expected to provide. In addition, membership criteria could be sufficiently flexible to allow representation by entities that might receive grants from QNRF (providing valuable "customer feedback"), whereas these entities would be precluded from board membership because of conflicts of interest.[1] Finally, a steering committee would consist mainly of people affiliated with QF, institutions in Education City, and Qatar University, with relatively little representation from outside, whereas a board would consist almost entirely of people from outside Education City.

Creating an Organizational Structure for QNRF and Hiring Staff

The original QNRF business and implementation plans called for a small start-up team. At the time RQPI began to work with QNRF in 2006, QF, using its leadership's professional contacts, hired a team of four people. A start-up director would oversee QNRF's launch; a special projects officer would direct the first program, the UREP; and two program managers would supplement their efforts, deal with two important sectors chosen by QF (health and biosciences, and industry and engineering), and assist with starting the NPRP. These four individuals had extensive responsibilities. To meet the many needs of QNRF, they needed a more-developed organizational structure and more staff.

The original QNRF business and implementation plans approved by QF in 2004 included a very high-level organizational structure and job descriptions. This structure outlined the framework for basic roles within the organization, but it lacked the level of detail necessary to staff QNRF fully. Therefore, RQPI worked with QF to devise a more-detailed organizational structure that developed and went beyond the original conception (Figure 2.1). A director, reporting initially to the steering committee and then later to a governing board, would lead the organization. He or she would be responsible for oversight of QNRF, with staff divided

[1] Education City and Qatar University each chose one representative for a committee seat. In informal discussions, university leaders indicated that they considered representation important during the start-up and viewed conflict of interest as manageable because of the strong working relationships in the university community.

Figure 2.1
Initial QNRF Organizational Structure

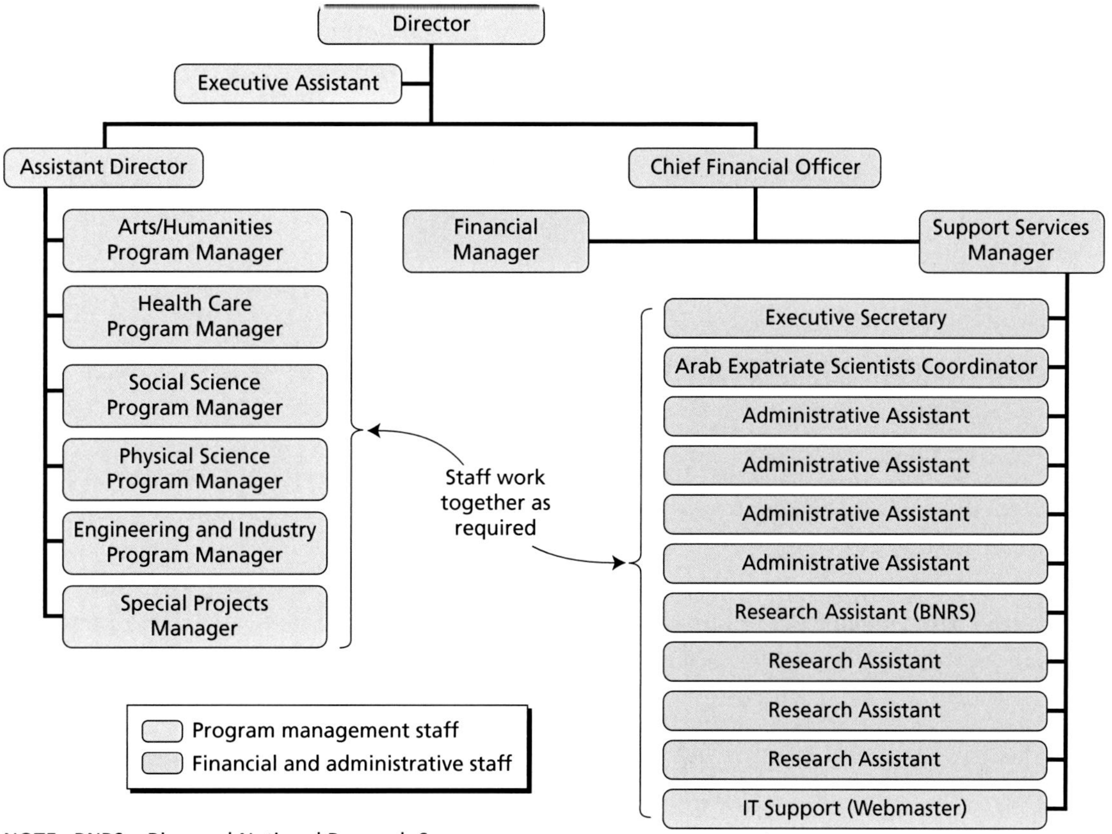

NOTE: BNRS = Biannual National Research Survey.

RAND *TR722-2.1*

into two categories: first, program management, and second, financial, administrative, and research-assistance staff. An assistant director would supervise the program managers. One program manager would manage grants in each general research field, or "bin." The bins consisted of arts and humanities, health and biosciences, social sciences, physical sciences, and energy and industry. (The bin system, which is used for both the UREP and the NPRP, is discussed in Chapter Six.)

A chief financial officer (CFO) would supervise a financial manager, a support-services manager, and the rest of the administrative staff. The administrative staff and research assistants, in turn, would jointly support the program managers as needed. The support-services manager would be in charge of allocating the time of the administrative staff and research assistants, in coordination with the program managers. The support-services manager would also supervise the work of the Arab Expatriate Scientists (AES) coordinator, who would facilitate a special project assigned by QF to create a network of distinguished scientists originally from the Middle East who were working in the United States, Europe, and elsewhere.

This structure was intended to make QNRF a "lean and agile" organization, limiting the total number of staff needed and creating a matrix structure, in which an employee could work on different projects or report to different people at different times (Ford and Randolph, 1992; Burns and Wholey, 1993). With this organizational structure, staff members could be used efficiently, taking on different roles depending on the needs of the organization at various times. For example, rather than having their own teams of support staff, program managers would share access to a pool of research assistants, administrative assistants, and coordinators, each of whom would provide various kinds of support functions. The support-services manager would allocate access to the support staff's time, distributing staff effectively, according to need. This would allow QNRF to best deal with the ebbs and surges of work during funding cycles. Having common processes across the organization would enable support staff to work on several different teams. The flexibility of the matrix structure thus proved important in enabling a small number of people to take on a large array of responsibilities in the start-up phase.

Lessons Learned from Building QNRF's Governance and Management Structure

The QNRF-RQPI team that determined the initial governance and administrative structure faced challenges and adapted on the basis of experience over time. For example, QNRF was understaffed during much of its start-up, for several reasons. To begin with, the organization expanded its scale quickly without adding enough new staff to keep up with increased demands. This placed a heavy burden on an overstretched team. Hiring and retention were also a challenge for QNRF, because of a lack of qualified people in Qatar, in addition to the long hours and flexibility required for working in a start-up organization. Internal processes were not able to keep up with the hiring needs, and only over time were more staff members hired. The QNRF-RQPI team realized that even the expanded "lean and agile management structure" was insufficient to meet QNRF program needs; the initial design was too lean. QNRF staff recognized that they needed a plan to revise and expand the organization's structure after its start-up in order to keep pace with its responsibilities. In particular, QNRF recognized the need for new program managers for public policy and management science and computer science and information technology (IT), along with a variety of additional staff.

In addition, although the steering committee was created as an interim body to oversee the start-up, it became clear that a steering committee would remain useful even after a governing board was put in place. As noted earlier, a steering committee is flexible, could meet regularly on a monthly basis, and could make decisions more quickly than a governing board, which would meet less frequently. Composed mainly of representatives of institutions located in Education City, a QNRF steering committee would also have local knowledge that could inform decisionmaking. RQPI recommended that the most effective governance model would be a combination of a steering committee for mid-level decisions and a governing board for higher-level decisions and overall program direction.

CHAPTER THREE

Establishing Guiding Principles for the Design of QNRF's Research Programs

After the initial QNRF governance structure was established, it was necessary to design QNRF's programs. The original QNRF business and implementation plans had established initial concepts for the programs, and the QNRF-RQPI team decided to revisit them, for two reasons. First, the concepts had been outlined at a very high level and needed more-concrete plans to make them operational. Second, because the QNRF business and implementation plans had been written several years earlier, in RQPI's 2004 study, the concepts had to be reviewed to make the programs appropriate for the rapidly evolving circumstances of Qatar's academic and research community.

Before the QNRF-RQPI team designed QNRF's first research programs, it established six principles to use as guidelines:

1. QNRF programs should aim to create a research culture in Qatar.
2. Program designs should include attractive incentives.
3. Programs should have one set of policies that can accommodate research in different parts of the world.
4. Programs require "buy-in" from participating institutions.
5. QNRF should learn from its own experience in designing programs.
6. Policies should be clear, transparent, and consistently applied.

The QNRF-RQPI team developed these principles jointly through a series of team meetings. Some of the principles were based on the vision, mission, and goals in the QNRF business and implementation plans, while others were developed on the basis of input from and interviews with stakeholders (including officials at research institutions in Qatar) about characteristics that are important in a granting institution. The principles are described below. Examples of how the principles were applied are given in Chapters Four, Five, and Six.

QNRF Programs Should Aim to Create a Research Culture in Qatar

The original QNRF business and implementation plans listed three goals QNRF would need to pursue to foster a culture of research in Qatar:

1. Build human capital in Qatar.
2. Fund research that will directly address problems of interest to Qatar, the region, or the world.
3. Raise Qatar's international profile in research.

The QNRF-RQPI team took as a core premise that the design of the QNRF research programs should contain elements that facilitate progress on all of these goals.

Build Human Capital in Qatar

Qatar has great wealth in natural resources and physical capital, but it faces challenges in building human capital. It has a small population that is just starting to take advantage of recent reforms in the education sector. The nation is undergoing a comprehensive restructuring of its K–12 education system to produce more graduates who have the skills to continue their studies at the university and postgraduate level and to fill the many professional needs of the society. Universities worldwide must be able to attract foreigners to fill faculty positions, in addition to developing local faculty. Research grants and related activities (e.g., conferences and comprehensive surveys of research) can do much to attract, retain, and develop faculty at Qatar University and Education City or researchers in other organizations.

Fund Research That Will Directly Address Problems of Interest to Qatar, the Region, or the World

Qatar wants much of its research funding to go toward projects that will benefit its society, diversify its economy, or address key research questions of importance to the world.

Raise Qatar's International Profile in Research

Raising Qatar's international profile in research is not simply aimed at building prestige. Building an international profile could diminish the geographical isolation of researchers in Qatar by helping them integrate with the global research community—to both their own benefit and that of the country. Researchers in Qatar will be better able to advance their own careers and make significant contributions in their fields if they are actively collaborating with other researchers around the world, taking advantage of the knowledge and skills gained through partnerships.

Program Designs Should Include Attractive Incentives

Incentives play an important role in creating both the motivation for research and a supportive research environment. The QNRF-RQPI team assumed that the most fundamental motives for performing research would be intellectual curiosity and a desire to explore topics of interest to the researchers themselves, their universities, their academic fields, Qatar, the region, or the world. But other incentives are required as well. The QNRF-RQPI team interviewed deans and faculty at the universities in Qatar to hear their opinions about incentives that would motivate faculty to engage in research and prompt universities to support them.

For individual researchers, these incentives included

- Publications and academic advancement through research accomplishments
- A staff that could assist with research
- The freedom to follow research interests
- Time off from heavy teaching responsibilities and other duties to spend on research
- Promotions and recognition
- Adequate facilities

- Extra pay in return for extra effort
- Travel.

For institutions, the interviewees cited

- The opportunity to collaborate with institutions outside of Qatar
- Appropriate freedom to manage their own budgets
- Funding mechanisms for hiring and importing graduate students
- Clear policies on intellectual-property (IP) rights that preserve incentives for researchers and institutions to pursue research
- Options for using reasonable portions of QNRF grants for indirect costs.

Programs Should Have One Set of Policies That Can Accommodate Research in Different Parts of the World

QNRF grant programs were intended to involve various organizations with different kinds of policies, procedures, and administration; in different phases of maturity; and in different parts of the world. Grant-funded research is very new for institutions in Qatar. Although many faculty members in Qatar had significant research experience elsewhere (in the United States, Europe, or the Middle East, for example), Qatar University as an institution had had very little experience with administering outside grants and conducting research, although it had recently begun several small research programs for its faculty. Education City universities faced similar circumstances. Whereas the Education City universities were branches of research institutions in developed economies with faculty experienced in research and well-articulated institutional research policies, they needed to develop research policies and administrative infrastructure unique to the circumstances in Qatar, while also adhering to home-campus policies and procedures. Consequently, QNRF programs sought a single set of policies that could fit the developing research administrative infrastructure at Qatar University and Education City, as well as the mature research systems in the United States, Europe, and elsewhere.

With this need in mind, the QNRF-RQPI team took as a point of departure research policies inspired by U.S. institutions such as NSF and the National Institutes of Health (NIH) (National Science Foundation, 2006; National Institutes of Health, undated). It also acknowledged that this model would need to be adapted to circumstances in Qatar, taking into account what was most appropriate given the research goals, opportunities, and constraints in Qatar, as well as feedback from Qatar's research institutions. In this way, it could devise policies that would work for all of the institutions involved.

Programs Require "Buy-In" from Participating Institutions

The QNRF-RQPI team decided that the QNRF programs should incorporate direct input from universities and other research organizations in Qatar that might eventually provide the human capital and facilities for QNRF-funded research projects. The team believed that direct contact with these institutions during the design phase would illuminate the needs of Qatar's emerging research community and would call attention to potential problems that might limit an institution's participation. These issues could then be addressed in the programs' design. Feedback from the institutions and how it influenced policymaking are discussed Chapters Four, Five, and Six.

QNRF Should Learn from Its Own Experience in Designing Programs

Experimentation and continuous learning from experience can provide valuable lessons that can be applied in real time to significantly improve programs. The QNRF-RQPI team determined that the design of QNRF programs should take into account ways to build upon each other as QNRF develops. For example, smaller, lower-profile programs could be used as a means of learning by trial and error, with QNRF staff and the planning team thinking through lessons learned in design, administration, incentives, peer review, and communication and later applying them to larger programs. Also, the earliest funding cycles of new programs could be viewed as learning cycles, enabling the programs' basic infrastructure to be deepened and improved upon in subsequent cycles. The following chapters discuss lessons learned and how QNRF changed its policies or procedures on the basis of those lessons.

Policies Should Be Clear, Transparent, and Consistently Applied

As a new organization that would give research grants on the basis of merit, QNRF needed to earn the trust of outside stakeholders. One way of gaining this trust would be to create clearly stated policies for awarding the grants and to apply them in a transparent and consistent way. Clearly stated policies and procedures would ensure that QNRF staff and stakeholders were all "on the same page." The QNRF-RQPI team therefore published the QNRF policies in the RFPs for the research programs.

Applying the set of principles it had defined, the QNRF-RQPI team developed working designs for QNRF's first two grant programs, the UREP and the NPRP.

CHAPTER FOUR

Planning and Launching the Undergraduate Research Experience Program: The First Three Funding Cycles

In October 2006, a few months after QNRF was created, it launched the UREP in order to produce results that could be demonstrated right away, even if that meant starting programs before the organization was completely structured. The decision to make the UREP QNRF's first program—before unveiling the larger program, the NPRP—stemmed directly from the guiding principle that QNRF should develop its grant programs in a way that would ensure that they could build on each other, with later programs incorporating lessons learned from earlier experiences. In addition, the UREP was a model of the NPRP on a smaller scale: Many of the UREP processes—such as developing an RFP and organizing peer reviews—would be similar to those needed for the NPRP. In this way, QNRF could apply lessons learned and be more effective than it would be if it had started the major program "cold."

Designing the UREP

In designing the UREP, the QNRF-RQPI team's primary goal was building human capital.[1] Students constitute Qatar's future workforce, and the team thought that it should stimulate an array of research opportunities for that important pool of human talent. Research projects funded under the UREP would be led by university faculty, who would direct the work of one or more students. Any undergraduate institution in Qatar, including Qatar University, the College of the North Atlantic, and the institutions in Education City, could host a UREP project. Promoting learning by doing, the UREP would encourage mentorship by faculty. The QNRF-RQPI team assumed that a research culture is best communicated via hands-on, research-team endeavors as a supplement to the normal classroom activities and that mentorship is an important dimension of the process.

QNRF planned to solicit proposals from Qatar's academic institutions in each funding cycle.[2] Proposals could involve either one academic department or a group of departments. They would specify possible undergraduate research projects and list faculty and other researchers who would manage those projects. The lead faculty member receiving the award would be responsible for selecting undergraduate participants and research topics and would then take the principal role in managing the students. Most individual projects would be undertaken within the academic departments receiving the UREP awards. However, collaborative proj-

[1] For an overview of undergraduate research programs elsewhere in the world, see National Science Foundation, 2006, 2010, 2011; Lopatto, 2007; University of Pittsburgh, undated; and Nagda, 1998.

[2] The text of the UREP's inaugural RFP is reproduced in Appendix A.

ects, in which faculty-mentored undergraduate research could involve commercial firms and government offices in some way, would also be welcomed. Projects would not be limited to the academic year but could take place during semester and summer breaks, giving students the opportunity to do research on an ongoing basis. The grants would provide $4,000 for the student, $2,000 for the faculty mentor, and $4,000 for equipment and travel (for example, to conferences or for fieldwork).

The specifics of the QNRF-RQPI team's design for the UREP were intended to enable undergraduates in Qatar to gain experience in conducting high-level research above and beyond what they would get through ordinary coursework during the academic year. Students attending universities in Education City or Qatar University who participated in the program would be more likely to receive a higher-quality education and a firmer foundation in research than they would otherwise. They would gain valuable practical experience in team-based collaboration with faculty, postdoctoral fellows, graduate students, undergraduates, and other research staff in Qatar. Research would assume a greater value in their eyes. The experience might make some more likely to continue on to graduate school, and all of this would provide them with a special type of workforce training (Lopatto, 2007).

While students were the primary focus in designing the UREP, the program was also intended to build human capital in other ways. The team anticipated that UREP awards would make faculty more aware of the possibility of conducting funded research in Qatar—which is particularly important in its own right and was also a way of creating a more favorable environment for the eventual launch of the NPRP.

Human capital can also be built by creating an institutional memory and a body of work. Accordingly, the QNRF-RQPI team integrated requirements into the UREP program design specifying that the results of student research projects must be archived and published,[3] at least in electronic form. These requirements were aimed at ensuring a public record of both projects and student researchers. This record would help publicize research done under the UREP and would also provide an archive of work that future researchers might draw on to pursue follow-on studies. In addition, the archive could be used to evaluate QNRF's impact on Qatar's development over the long term.

A key step in designing the UREP was creating the initial draft of the RFP that would be sent to Qatar's academic institutions. The RFP would give QNRF an opportunity to get buy-in from those institutions, in line with one of its guiding principles. After an initial draft went through several rounds of revisions by both QNRF and RQPI staff, the QNRF-RQPI team distributed the revised draft to the deans of the branch campuses in Education City, the colleges at Qatar University, and the College of the North Atlantic. Team members then held individual meetings with the deans to explain the RFP and solicit further feedback and suggestions for improvement.

Giving the deans the opportunity to participate in the program design produced a number of benefits. The face-to-face meetings advertised the programs and put institutions on notice that they could participate. Using input from the universities gave them a greater sense of ownership and stake in the program. At the same time, it gave the team insight into the needs of Qatar's research community and into potential stumbling blocks, allowing it to tailor the UREP design to the circumstances of key stakeholders in the community.

[3] Most such projects would be documented in student reports.

Some of the interviewees suggested that faculty be awarded some funding in recognition of the effort associated with mentoring students. A financial incentive might provide busy faculty, many of whom have heavy teaching loads, with more inclination to participate. In response to this suggestion, the team incorporated into the UREP program design a standard faculty award of $2,000 per student.[4]

The UREP selection process was designed to operate on the basis of merit. The QNRF-RQPI team recognized that a peer-review system is the key to assuring that high-quality proposals get funded. With this in mind, RQPI staff provided QNRF with the names of potential peer reviewers, drawing from professional contacts in the Middle East and the United States, as well as from a list provided by NSF. The team also drafted peer-review guidelines that provided referees with a scoring system to use in evaluating UREP proposals. The evaluation criteria, matching the point system to the goals of the program, were developed in a series of meetings. The scores were to be distributed as follows:

- 30 points for anticipated benefit to the undergraduate student
- 20 points for anticipated benefit and relevance of the proposed activity to Qatar
- 20 points for intellectual merit of the proposed activity
- 30 points for the plan for mentoring and oversight.

The description of the point distribution was meant to be broad enough that reviewers could apply the same criteria to proposals in fields as diverse as the arts and robotics.

Results of the UREP's First Funding Cycle

The UREP's first RFP attracted submissions from every college and university in Qatar. A total of 120 proposals were received from faculty at Qatar University and Qatar's branch campuses of Texas A&M, Weill Cornell Medical College, Georgetown University, Carnegie Mellon University, Virginia Commonwealth University, and the College of the North Atlantic. This was several times the projected response rate originally envisaged in the QNRF business plan for the UREP's first year. The extent to which the actual response exceeded the projections was perhaps due to the interest and intellectual engagement of faculty and students in Qatar, the financial incentives that the UREP design provided to them, and a concerted effort by RQPI and QNRF staff to provide information about the program to potentially interested applicants. Not only did developing and running the UREP program provide valuable experience for the QNRF-RQPI team to use in developing the NPRP, it also demonstrated that QNRF was a viable grant-making institution, at least on a small scale.

QNRF awarded $1,322,000 in grants in the UREP's first funding cycle. All of the universities with faculty who submitted proposals won grants. More than half the 120 proposals received (61 in total) were judged to be of sufficient quality to receive a UREP grant in the first round.

---

[4] It is difficult to gauge the incentive effect of the award on faculty in Education City branch campuses because contractual agreements with their home institutions in the United States and Europe prevent many of them from accepting the award directly as a cash payment. In such cases, the award is deposited in a university account, where it can be drawn on for research-related expenses.

About 20 percent of the applications were declined on administrative grounds. This was an important indication to QNRF staff that some of the faculty in Qatar may have lacked experience in proposal writing. In keeping with the principle of applying lessons learned to future funding cycles, QNRF decided to provide training sessions on how to write successful proposals in subsequent cycles.

Among the most important indicators the QNRF-RQPI team tracked during the UREP's first round of funding was the extent to which the proposals submitted and awarded met the program's objectives. The UREP's core objective is to give undergraduate students an opportunity to interact directly with professors in a research setting, in addition to the interactions that take place in the classroom environment. The large majority of the proposals funded in the first funding cycle involved student-to-professor ratios of 2:1 and smaller (Figure 4.1), meaning that the majority of the proposals funded would provide opportunities for direct mentorship and close professional contact between students and professors.

QNRF aspires to fund research in many disciplines in its programs. Initially, the QNRF-RQPI team was concerned that topics in engineering would dominate the UREP proposals, because of the obvious importance to Qatar of successful research in those areas and the large engineering departments at some universities in Qatar. In fact, this concern proved to be unfounded. About three-quarters of the UREP proposals submitted in the first funding cycle were in the physical sciences, health, arts and humanities, and social sciences (Figure 4.2). (The bin system was developed later for the NPRP, but the results below are divided among what later became the bins. Additional bins, such as in public policy and management, were added later.)

The UREP also offered an opportunity to promote and increase the participation of Qatari women in the country's workforce. The RQPI-QNRF team had anticipated that the availability of opportunities to conduct research as undergraduates might encourage more of Qatar's women to pursue graduate studies or increase their skills as they entered the labor

Figure 4.1
Student-to-Professor Ratios in the First Cycle of UREP Funding

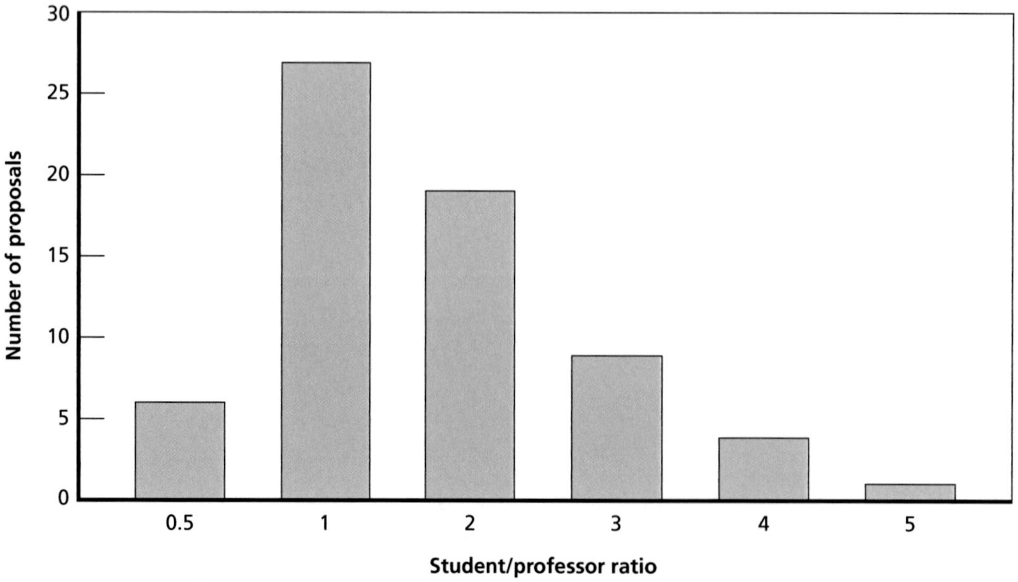

RAND *TR722-4.1*

Figure 4.2
UREP Proposals in the First Funding Cycle, by Discipline

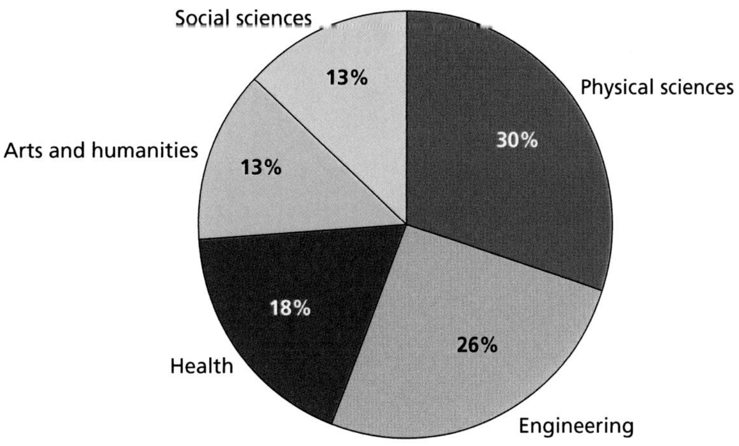

RAND *TR722-4.2*

force. The awarded UREP grants were encouraging in this regard: Twice as many grants went to research teams comprising only female students as went to teams comprising only male students (Figure 4.3). Gender was not a selection criterion; the grants were based on merit. The ratio of female to male teams submitting proposals was roughly the same. This raised concerns about male participation in the program. However, seven of the awarded proposals were submitted by teams of mixed gender, which was interesting, since male and female students at universities in Qatar traditionally do not collaborate in their work because of cultural norms. These mixed teams might have consisted of either Qatari students who were more comfortable with working with the opposite gender or non-Qatari students who were not subject to the same norms. However, QNRF did not collect data on nationality during the first cycles, so these suppositions could not be verified.

Figure 4.3
Gender Composition of Research Teams Whose Proposals Were Approved

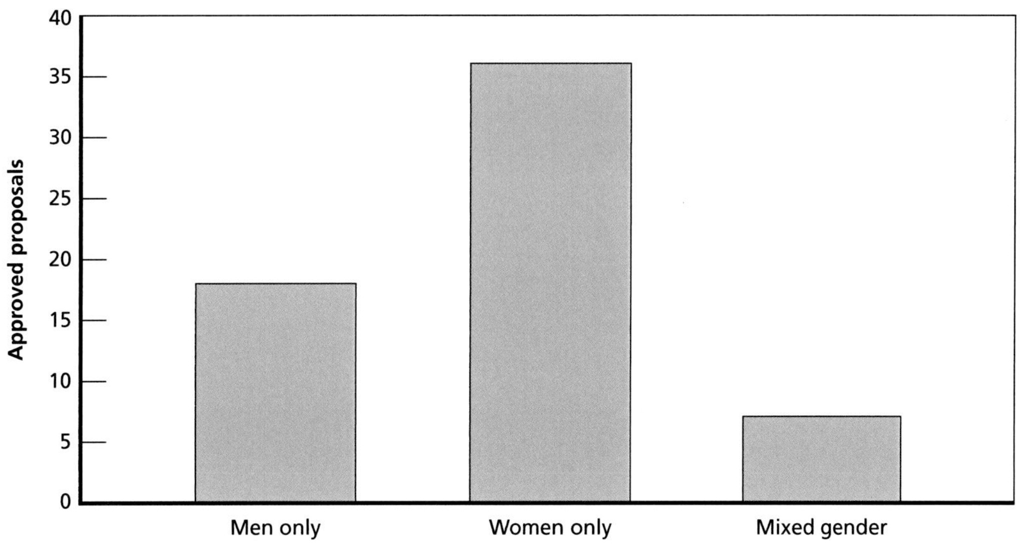

RAND *TR722-4.3*

Results of the UREP's Second and Third Funding Cycles

The UREP's second RFP went out about six months after the first, in March 2007. A total of 86 proposals were received—considerably fewer than in the first cycle. However, the second group of proposals was generally of higher quality than those in the first round, as reflected in the average peer-review scores. Less than 5 percent of the 86 proposals were declined on administrative grounds, in contrast to the nearly 20 percent in the first cycle. Eighty-two of the 86 proposals went on to peer review.

Reviewers gave the second group of proposals a mean score of 75.5 points, nearly five points higher than the scores of those in the first cycle. In light of this improvement, QNRF raised the cut-off score for funding in the second cycle from 70 to 75 points. Even with this more-stringent requirement, the proposal success rate—i.e., the number of proposals awarded as a percentage of those submitted—increased from 50 percent to 62 percent. The ratios of faculty to students, the split of grants among research areas, and the participation of women were largely similar to those in the first cycle. The UREP awarded $1,436,500 in funding and honoraria in the second round, an increase of $114,500 over the total awarded in the first round.

Proposals for the third cycle were due in October 2007, and QNRF awarded $1.4 million in December 2007. A total of 96 proposals were submitted in the third cycle. Of these, 87 were forwarded to peer review, and 9.3 percent failed to pass administrative review. (The difference between the rates in the two cycles was perhaps due to the small sample sizes.) The mean score rose by 1.1 points in the third cycle, to 76.6 points.

The increase in the average scores of proposals over the three cycles, as well as the corresponding decline in the proportion of proposals rejected due to administrative problems, was promising. These changes may have come about because of (1) learning and a general improvement in the quality of proposals, (2) faculty and administration becoming more selective about the proposals presented to QNRF (i.e., exercising greater quality control), or (3) a combination of the two. QNRF's training sessions on proposal writing may also have contributed to the change in rates.

Table 4.1 summarizes the data on the proposals in the first three cycles of the UREP.

Table 4.1
Proposals in the First Three UREP Cycles

	First Cycle	Second Cycle	Third Cycle
Number of proposals received	120	86	96
Percentage rejected on administrative grounds	20	5	9
Mean score of reviewed proposals	71	75.5	76.6
Cut-off score for funding	70	75	75
Total number of awards	61	53	42
Proposal success rate (%)	50	62	44
Total amount awarded ($)	1,322,000	1,436,500	1,400,000

Applying Lessons Learned from Earlier Experiences to the UREP's Second and Third Funding Cycles

The QNRF-RQPI team viewed the UREP's first three cycles as opportunities to gauge the demand for funding and the supply of high-quality proposals. They decided that the cut-off score after the third cycle would be 80, meaning that UREP funding might change from year to year.

QNRF retained most of the UREP's original policies and procedures in the second cycle. Only two significant changes were made. First, to promote quality individual attention and close professional interaction among the faculty mentors and the student researchers, a cap was placed on the number of students that any one faculty member could mentor. The maximum ratio became 3:1. Most of the projects already fell into that range, but a limit was established for those that did not. Second, since more QNRF program managers had been hired, QNRF's director distributed the process of assigning peer reviewers for the UREP among them. QNRF could now harness the subject-matter expertise of the program managers to facilitate the process of identifying and inviting reviewers.

QNRF changed the scoring mechanism for the third cycle. By the time that cycle began, QNRF had launched the NPRP,[5] which had a scoring system to evaluate proposals that was different from the system used in the UREP's first two cycles. Whereas the UREP had used a simple arithmetic mean of the scores—a sum-of-means approach—a sum-of-medians approach was used for the NPRP, taking the median score of each evaluation criterion and summing the scores. The early experience of the NPRP showed that the median was more robust with respect to outlier data than the simple arithmetic mean.[6] Consequently, in accordance with the principle of continuous learning to improve program design, QNRF's leadership switched to a sum-of-medians score in the UREP's third cycle.

After three cycles, the QNRF staff continued administering the UREP without RQPI support. Two cycles of the UREP would be run per year after that.

---

[5] Although the UREP's third cycle took place after the NPRP's first cycle, we discuss it here to show more clearly the developments that took place over the course of the three funding rounds.

[6] This is discussed in greater detail in Chapter Seven.

CHAPTER FIVE

Laying the Groundwork for the National Priorities Research Program: Designing Policies for the Program and Developing an Initial Request for Proposals

Once QNRF had experience to draw upon from the UREP and its policy infrastructure was in place, it turned its attention to the NPRP, which would be its primary tool for supporting larger research projects in Qatar. The QNRF-RQPI team undertook a significant amount of planning and worked through key decisions, the most important of which concerned the structure of the program (the research-funding model and the number and size of the grants), the research priorities of the program, design features that would meet the priorities, IP policies, and administration of a peer-review process.

The team also designed the RFP, which would be the most important policy document of the program. The RFP was to present policies and processes in a clear and transparent fashion for both QNRF staff and the research organizations in Qatar that would apply for grants. The QNRF-RQPI team considered the first cycle of the NPRP as a learning cycle, in which the basic infrastructure of the program would be set up; it would then be deepened and improved upon in each later cycle, on the basis of experience gained along the way.

Designing the Structure of the NPRP

The concept for the NPRP in QNRF's initial business and implementation plans involved the funding of only a limited number of large grants—a model in which QNRF laid out research priorities. Each major university in Qatar was to have one multi-investigator, multidisciplinary project that might also entail collaboration with other institutions inside or outside of Qatar. The program was to accept no more than 16 proposals.

However, on the basis of interviews with research stakeholders, QNRF-RQPI team analysis, and experience with the UREP in its first two funding cycles, QNRF decided that a different approach was in order. Involving many researchers with creative ideas and judging merit through a competitive process worked with the UREP, and the QNRF-RQPI team believed that it would be more effective to adopt a complex, adaptive research-funding model in which QNRF gave grants to individual projects based on the merit of the projects.[1] Many grants, rather than fewer, might lead to a more dynamic research community in Qatar. But awarding many grants might result in no critical mass developing in any research area because the funding was too widely dispersed.

Rather than attempting to organize a few large-scale projects at major educational institutions in Qatar, QF and QNRF decided (at RQPI's recommendations) that they would issue

[1] For a discussion of complex adaptive systems in organizations, see Anderson, 1999.

a general RFP to all research organizations in Qatar and would then make the final grant decisions through a competitive process based on clearly defined criteria.[2] Any organization in Qatar, including Qatar University, the Education City branch campuses, the College of the North Atlantic, Hamad Medical Corporation, private companies, and government entities, could apply for grants. However, opening up the program to multiple research proposals and a competition would entail a very different level of effort than that of the original plan. The effort would entail determining and stating in advance what the research priorities should be, how to focus funding to meet them, and how to administer an endeavor that would potentially involve many more proposals and peer reviewers than originally envisioned.

Thinking Through Research Priorities

Once the decision had been made to award multiple grants on a competitive basis, the question arose as to how to choose research areas for the grants. During initial planning for the NPRP in the QNRF business and implementation plans, it was assumed that QNRF would first select specific research areas that were key to Qatar's national interests and would then concentrate its resources on those areas in a certain order of priority. As planning progressed, however, it became clear to the QNRF-RQPI team that this approach would present a number of challenges. There were three main concerns:

1. It would be very time consuming to answer well the question of what Qatar's priority research areas should be.
2. Limiting funding to specific research areas would be politically sensitive.
3. Focusing research funding on particular areas could potentially lower the quality of research if Qatar did not have researchers with the requisite skills for those areas.

Determining Qatar's Research Priorities

At the time the NPRP was being designed, Qatar had not formally articulated a national policy on priorities for research. Although QF had commissioned some preliminary work by Angle Technology (an international management services firm with expertise in science and technology policy) to determine what research areas might be important to Qatar, this work had no official standing. Moreover, the Angle Technology recommendations were limited to a few research areas that did not completely involve the branch campuses in which QF had invested. QNRF decided that it would support a broader spectrum of activities to achieve the goal of building human capacity and research culture. QNRF wanted to ensure that funding was available for the specialties of the universities and research organizations that were already present in Qatar. Because Qatar had made a strategic decision to bring in branch campuses that focused on key areas of research, it would be important to ensure that QNRF programs provided a research-funding infrastructure that supported their faculties. QNRF did not want to create a program for which many research organizations in Qatar would be ineligible.

---

[2] For a discussion of recommending an appropriate funding mechanism for QNRF, see Greenfield et al., 2008. Also see Alfred P. Sloan Foundation, 2008; DARPA, undated b; NIH, 2011; and NSF, 2009.

Political Sensitivity of Limiting Funding to Specific Research Topics

Setting priorities for research meant deciding what would not be funded, as well as what would be funded. Meetings with stakeholders demonstrated the political sensitivity of specific research priorities. Very rarely could a group of stakeholders agree on a definitive list, as various participants in the process expressed concern that topics they deemed important were not included. In some cases, limiting the list of priorities to those topics that stakeholders considered essential would have worked against Qatar's goal of diversifying the economy, since QNRF wanted to invest in new kinds of research. It became apparent to the QNRF-RQPI planning team that disagreement over an exclusive list of research topics could indefinitely delay the start of QNRF's programs.

The Potential Effect on the Quality of Research of Focusing Funding on Particular Topics

Qatar was only starting to build its research culture and capabilities. Consequently, there might not be researchers in Qatar with the necessary expertise for some research topics. The QNRF-RQPI team recognized that requiring QNRF funding to be used for research on predefined topics was likely not to yield strong results if priorities did not match capabilities. In addition, limiting research funding to only certain topics might constrain the ability of universities in Qatar to attract and retain the best researchers in areas for which there was no funding available.

The intractability of these challenges called for a flexible approach. Therefore, in the interests of getting an effective program started quickly, RQPI recommended that QNRF might more productively think in terms of building capabilities rather than pursuing specific topics. Instead of focusing on assigning research topics a certain order of priority, the NPRP should use the three QNRF goals—building human capital; funding research in the interest of Qatar, the region, or the world; and raising Qatar's profile in the international research community—as guidelines for selecting topics to fund.

Because the first priority among these three goals was building human capital in Qatar, RQPI recommended that this be the NPRP's initial emphasis. Funded research would need to improve the quality or enhance the size of Qatar's workforce in some way. Thus, the NPRP's policies were to be decided with the following three objectives:

- To attract, develop, and retain faculty at Qatar University and Education City and researchers at other institutions in Qatar
- To provide incentives to build an institutional infrastructure in Qatar that supports research
- To advance knowledge and technology transfer within Qatar through collaboration with institutions outside of Qatar.

With RQPI's recommendation, the QF leadership decided that the best way to achieve these objectives was to allow researchers in Qatar to propose topics of their own. The topics submitted would then be evaluated by international, independent peer reviewers on the basis of criteria laid out by QNRF. In later cycles of the NPRP, QNRF could consider focusing its support on research of special interest to Qatar.

Since applicants would be able to propose topics of their own choosing in any research field, questions quickly arose about how QNRF would allocate its funding. Would it opt to award certain amounts to any particular field of research? Would all proposals, regardless of their research field, compete using the same point system? What if different research fields were to yield different standards for peer-reviewer rating of proposals?

To address these questions, the QNRF-RQPI team decided to specify in the RFP that when an applicant submitted a proposal, he or she must chose a bin (i.e., a research discipline) in which the proposal would compete. After all of the proposals had been submitted and peer-reviewed, the QNRF steering committee could choose different levels of spending within the various bins to balance out conflicting opinions about how NPRP funds should be allocated across research fields. The steering committee would not attempt to decide how QNRF would allocate grant money across research fields before any proposals had actually been received.

The bins, decided collaboratively by the QNRF-RQPI team, were

- Industry and engineering
- Physical sciences
- Arts and humanities
- Social sciences
- Public policy and management science
- Computer science and IT
- Health and biosciences.

The QNRF-RQPI team considered this selection of bins to be broad enough to accommodate proposals in any major research area and, in particular, in areas most likely to be proposed by the research institutions in Qatar. The bins were created for ease of administration, so that proposals could be sorted and managed. Program officers could be hired to manage the various bins.

Developing an Intellectual Property Policy

Potential recipients applying for grants would need to know what rights they would have to the ownership of IP they might produce through research funded by QNRF, as well as to the revenue stream that IP might generate. The IP policy would govern not only NPRP grants but also UREP grants.

A Complex IP Environment

QNRF's IP environment was very complicated, involving multiple competing factors:

- QF's interests in IP resulting from its grants
- IP policies of institutions in Qatar, such as Qatar University
- Home-campus IP policies of the U.S. branch campuses in Education City, some of which were subject to the particular laws of the U.S. states in which the universities were located
- Policies of other institutions around the world that would potentially collaborate in QNRF-funded grants
- Division of the ownership of IP rights among the researchers, the branch campuses in Qatar of U.S. universities, their home campuses in the United States, and QF.

This host of factors created difficult challenges for the QNRF-RQPI team. How would the IP rights be divided among the numerous interested parties, balancing short-term revenue interests for QF and QNRF and long-term interests of providing incentives for researchers and institutions to participate? Although QNRF would be funding the research and would want some benefits from any IP produced, if it retained too large a share in comparison with common practice worldwide, that could discourage researchers and institutions from undertaking research under QF's auspices.

The QNRF-RQPI team wanted QNRF's IP policy to achieve four goals:

- Create an environment that will encourage innovation by enabling researchers to maintain rights to their IP.
- Encourage participation of institutions in Qatar by making QNRF's IP policy compatible with the institutions' existing incentive structure.
- Motivate institutions with researchers who receive QNRF grants to establish management systems to support and facilitate the development and commercialization of IP.
- Generate revenue for QNRF or QF.

The IP Solution

To resolve these challenges, RQPI recommended the following policy.[3] First, for grants made to researchers at institutions with an existing IP policy, QNRF should adopt a compatible policy, with one exception: The home institution's share of revenues under its own IP policy would be split equally between QNRF and the institution.

Second, for grants made to researchers at institutions without an IP policy, QNRF should adopt the prevailing international model, with several modifications. The prevailing international model is currently in place throughout the United States[4] and in many nations in the European Union. In that model, the researcher's home institution owns the rights to the IP. Any revenue from that IP is split three ways—one-third goes to the researcher, one-third to the researcher's department or faculty,[5] and one-third to the researcher's institution. QNRF adopted a modified version of this model: The last third of the revenue stream is to be split between the researcher's home institution and QNRF.

Third, for grants involving other kinds of partnerships with various combinations of funding and award recipients (such as with the private sector), QNRF will negotiate separate IP and revenue agreements on a case-by-case basis, using the standing QNRF policy as guidance.

Developing the NPRP's Initial Request for Proposals

RQPI took the lead in developing the NPRP's RFP, working iteratively with QNRF staff. It was agreed that the design should be based on international best practices adapted to Qatar's

[3] See Greenfield et al., 2008, for a full discussion of RQPI's analysis of IP models around the world and references. See Mowery et al., 2001, for a discussion of more-defined IP incentives and their effect on IP development at universities in the United States.

[4] The prevailing international model became widely adopted in the United States in the 1980s.

[5] *Faculty* is defined as a department or group of departments dealing with a particular subject in a university or college. For institutions with a Qatar campus, we define *faculty* to mean the departments or group of departments in Qatar. This definition will be operative only in those cases where there is doubt over a particular institution's revenue model.

circumstances and needs. Accordingly, early in the development, RQPI conducted a comprehensive review of examples of RFPs and related processes (e.g., peer review) at funding institutions in the United States, including NSF,[6] NIH,[7] and the Department of Defense (DoD).[8] RQPI was primarily interested in the funding institutions in the United States because their systems would be compatible with those of many of the institutions funded in Qatar, such as those in Education City.

The QNRF-RQPI team used the iterative process of drafting the RFP to work through the policy and procedure decisions that were required for the program. In keeping with the guiding principles of clarity and transparency, the RFP was used to create an overview of the program to be used as reference for the QNRF staff and for the applicants.

As they did in developing the UREP's RFP, QNRF and RQPI staff interviewed the deans and other officials at Qatar University and the U.S. branch campuses in Education City. This was again in line with the guiding principle of securing the buy-in of institutions likely to participate in the program. Through these meetings, the QNRF-RQPI team was able to get a sense of the universities' readiness to participate in the NPRP. At the same time, team members encouraged the university representatives to share their recommendations and concerns about the program and invited them to comment on draft versions of the RFP.

Creating NPRP Design Features to Build Human Capital in Qatar

To ensure that NPRP projects would meet the first cycle's primary objective of building human capital and research capabilities in Qatar, the QNRF-RQPI team built requirements into the NPRP design to develop researchers already in Qatar, build institutional infrastructure, and encourage international collaboration. Seven of the key design features and the reasoning behind them are discussed below.

Encouraging Intellectual Freedom and Technical Merit in the Design of Projects

As noted above, the NPRP would initially fund high-quality research in whatever areas researchers in Qatar wanted to pursue, provided they could make the case that their research would build human capital and was of interest to Qatar, the region, or the world. The research community in Doha was still too new to focus on particular topics or limit what researchers could choose. Work that researchers in Doha wanted to undertake would be supported, helping provide the incentives needed to attract new researchers and retain those who were already there. The team believed that over time, this approach would strengthen and build the research community. Later, Qatar could choose to focus its research funding on particular topics, but limiting topics would not be useful for the first cycles.

Proposals would be ranked on a 100-point scale, with 75 possible points for the technical merit of the proposal, including original contribution to knowledge; significance of the research questions to the discipline; importance to Qatar, the region, or the world; organiza-

---

6 See National Science Foundation, undated, 2006, 2009, 2010, 2012.

7 See National Institutes of Health, undated; U.S. Department of Health & Human Services, updated February 24, 2012.

8 See DARPA, undated b.

tion, methods, and design; and evaluation plan. The other 25 possible points would be used to rank research-team qualifications, including qualifications of the principal investigators (PIs), team members, and facilities.

Requiring Researchers to Make Their Own Case for the Importance of Their Research to Qatar, the Region, or the World

While the focus for the first cycle was on building human capital, the QNRF-RQPI team was aware of the eventual need for QNRF to fund research that served either national, regional, or global interests. However, defining the kinds of research relevant to Qatar proved challenging. The team was also concerned that if research funding was limited to topics relevant only to Qatar, peer reviewers might interpret this as meaning that QNRF would support only research that specifically applied to Qatar. Theoretical research (research on health care, for example) that did not apply to Qatar might be disqualified for not being relevant in the short term, even though it might be very important to residents of Qatar and the rest of the world in the long term. The RFP therefore asked researchers to make their own case. This requirement was meant to be flexible enough to accommodate the diverse interests of researchers in Qatar. (The case made for the research topic was worth a small number of points in the peer-review process; the priority was funding high-quality proposals of interest to researchers in Qatar in order to build research capacity in the country.)

To stimulate the ideas of researchers who might apply for an NPRP grant, QNRF developed a list of sample topics to be included in the NPRP's first RFP. The list was based in part on the preliminary work by Angle Technology discussed above. It also included several new disciplines, in keeping with QNRF's mission of supporting research not only in applied science and technology but also in a range of other disciplines. The list was intended to serve as a source of ideas rather than to be exhaustive or exclusive. The NPRP's RFP openly encouraged researchers to submit proposals for topics not on the list as well.

Investing in Researchers in Qatar While Encouraging Collaboration with Researchers from Around the World

A key element of building research capability was collaboration with researchers in other parts of the world. Such collaboration would help build Qatar's reputation as a center of research, and raising Qatar's international profile would eventually play an important role in attracting and retaining skilled researchers. The RFP therefore laid out requirements for the locations of the research, including where the research was to be done and where funding was to be spent. Most of the research had to be done in Qatar, but a portion of the grant could be used by outside collaborators. The intent was to ensure that a substantial percentage of funded research, as well as the intellectual capital development derived from it, would take place in Qatar and would involve people who lived there—either Qataris or expatriates. At the same time, the RFP included the following provisions to attract participants outside of Qatar:

- At least 65 percent of a proposal's research budget would have to be managed by an institution in Qatar, allowing 35 percent to be managed by institutions outside of Qatar.
- At least 50 percent of the total person-hours spent doing the research would have to be spent in Qatar.

- At least one PI per project and a significant number of the research staff would have to reside in Qatar while the project was under way; the program also allowed co-PIs who were not resident in Qatar.
- Three bonus points (in addition to the 100 points the proposals would be judged on) would be awarded to proposals that involved partnerships with institutions in other countries.

Funding collaborative projects around the world is a unique feature of the NPRP; similar programs in other countries in the Middle East tend to limit grants to their own residents. Features encouraging collaboration were built into the program design to reduce the isolation of researchers in Qatar, to build new links between Qatar and the international research community, and to increase international recognition of research under way in Qatar.

Encouraging Collaboration Among Research Institutions in Qatar

The QNRF-RQPI team recognized that research collaboration in Qatar would also develop sustainable research capabilities within the country. To this end, the NPRP was designed to encourage projects that would partner researchers from different campuses in Qatar and would team academia with industry or the public sector in Qatar. Proposals that involved such collaborations would receive two bonus points (in addition to the 100 total points). The small number of bonus points involved might make a difference for only a few proposals at the margins; however, they served as a symbol that QNRF encouraged collaboration.

Funding Medium-Term Grants

The QF leadership decided that QNRF would fund grants ranging from $20,000 to $250,000 per proposal per year, according to the budget requests in the proposals that succeeded. They wished to fund research generously enough to attract interest from high-quality researchers, while stretching their budget to sponsor many researchers in Qatar. Conducting research in Qatar is more expensive than in other countries, because it often relies on graduate students and postdoctoral fellows at universities, and most universities in Qatar do not have graduate programs. Research projects would have to bring staff to Qatar and provide housing in one of the most expensive housing markets in the world.

QF also decided that grants should be for projects designed to take from one to three years to complete. QNRF hoped to fund the kind of sustainable, longer-term research that is necessary for building research capability and infrastructure. For projects running longer than one year, funding was to be available annually, after QNRF review of the project and a satisfactory progress report (defined as a research team's meeting the milestones set in the proposal). The emphasis on multi-year proposals was also in part intended to make QNRF more competitive with funding organizations in other countries, such as NSF, NIH, and the Defense Advanced Research Projects Agency (DARPA).

Incentivizing Research Institutions in Qatar to Establish Needed Policies and Infrastructure

The QNRF-RQPI team made a point of writing into the NPRP's RFP programmatic elements aimed at enabling institutions in Qatar that had never done research under the auspices of outside grants to do so. In this sense, research institutions were called upon to play a pivotal role in the proposal process.

The policies encouraged the involvement of a research office. Branch campuses in Education City and the College of the North Atlantic that did not have a research office would be encouraged to create one to support work with QNRF. They would then need to coordinate these new offices and policies with their home-campus research practices. Issues for the institutions to address included guidelines for dividing indirect costs between the United States and Qatar, ethical and legal obligations, finance and budget policies, and roles of the home institution and the Qatar branch campus in developing and implementing research policies. Qatar University has a research division, and QNRF's programs were meant to require its active involvement in the development of infrastructure and policies.

The RFP mandated that applications include two letters of support from the researcher's home institution: one from the relevant dean and another from the CFO (or the equivalent of either). These letters were to certify (1) that the participating institution had policies in place to support research (ranging from ethics and human-subjects protection to hiring research support staff and arranging for travel) and (2) that the institution verified that the proposal complied with those policies and the requirements of the RFP. The planning team intended for this to incentivize institutions to develop administrative policies supportive of research in Qatar and then monitor them. Having a dean or CFO sign off that a grant application was in line with university policies would ensure that responsible parties were coordinating the proposals and verifying that they followed university procedures. It would also make the dean or CFO responsible for creating the policies at the institution, if policies did not already exist, as well as for monitoring them. This would remove much of the burden and responsibility of monitoring the grants from QNRF itself. To further ensure that university research policies would be created, the QNRF-RQPI team encouraged the universities in Qatar that did not have research offices to create them.

Project budgets were a key issue. The letter of endorsement required from the CFO at the applicant's home institution had to certify that the budget conformed to the institution's approved financial policies on outside research funding, as well as to those of QNRF. This was intended to promote coordination among the budgets institutions were submitting.

Requiring a Letter of Intent

The NPRP design included a requirement for applicants to submit an online letter of intent that included abstracts of the proposed research before the proposals were due. QNRF hoped to use these letters to start locating suitable peer reviewers before the proposals were submitted. (The QNRF-RQPI team assumed that it would take a significant amount of effort and time to line up appropriate peer reviewers, particularly in the first cycle.) QNRF had a small database of reviewers that it had used for the UREP, but it would need many more who could review proposals on a wide array of topics. The QNRF-RQPI team did not know how many proposals to expect, so it relied on these letters to describe the topics that would be covered, as well as the numbers of peer reviewers that would be needed.

Creating a Process for Peer Review of Applications

In keeping with the guiding principle of transparency, the NPRP's RFP had to establish and lay out in detail the process by which proposals would be peer reviewed and decisions would be made about which grants to fund. The peer-review system had to be flexible enough to accommodate a great deal of uncertainty. It was not clear how many proposals would be submitted, how many proposals would be in each bin, or what the quality of the proposals would be. The QNRF-RQPI team anticipated proposals for a range of research topics spanning, for example, diabetes to robotics to Islamic studies. The peer-review process had to be robust enough to facilitate quality comparisons across proposals in many different fields. Because one of QNRF's goals was to help diversify Qatar's economy, the peer-review process had to ensure that a diverse group of research topics would receive grants and that they would not all go to research in one field, such as oil and gas, the mainstay of Qatar's economy.

The bin system proved very useful here. The QNRF-RQPI team integrated into the RFP the requirement that applicants must choose a bin in which the proposal would compete. This would make it much easier for QNRF staff to select peer reviewers; in addition, proposals in the same field would compete with each other within their bin rather than competing with proposals for all topics. The highest-scoring proposals within each bin, down to a decided cutoff point, could be chosen. QNRF would also have a way to prioritize subjects or to change the balance of the investment among bins if it so chose. It could decide to fund more studies in the arts, for example, and put more funding into that bin. The bins would provide a tool for managing changes in amounts invested in various research areas if QF chose to make such changes in the future.

QNRF also had to decide how to entice respected scholars around the world to serve as peer reviewers for this new organization located outside of the established peer-review networks in the United States and Europe, where many potential reviewers might be hesitant to invest their time. Researchers customarily agree to serve as peer reviewers within their own country without compensation, and they sometimes accept compensation from organizations outside of their country. QF and QNRF decided to provide compensation to peer reviewers to provide an incentive for highly respected researchers to participate.

CHAPTER SIX

Launching the National Priorities Research Program

The QNRF-RQPI team made efforts to ensure that the first NPRP cycle would attract high-quality proposals and that their review would be administered according to plan. After the release of the RFP in April 2007, QNRF communicated with the research community about the program and responded to their questions. As shown in Table 6.1, four months would be allowed for applicants to prepare their proposals. These four months would also allow QNRF to continue developing its own internal administrative infrastructure (including online submission processes) and policies. In July 2007, QNRF required letters of intent from applicants. These letters would give QNRF time to start finding peer reviewers—no trivial task given that QNRF would need nearly 1,000 reviewers (to meet the goal of five peer reviews per proposal) for the first cycle. Proposals were due in August. About 200 proposals were received—far more than the 16 the original QNRF business and implementation plans anticipated. QNRF and RQPI then scrambled to finish finding peer reviewers, coordinate the peer-review process, create policies about decisions on grants, and select projects to fund. In January 2008, awards were announced. This chapter describes the steps taken, results, and lessons learned in the first cycle.

Ensuring Publicly Available and Transparent Information for Applicants

QNRF wanted to make information about the program widely available. It also valued creating a reputation for fairness, consistency, and transparency in its policies and practices, in

Table 6.1
Time Line of the NPRP Launch

Date	Action
April 28, 2007	Program is launched and NPRP RFP is made publicly available.
Mid-May 2007	Three public presentation and Q&A sessions are held; Q&As from the sessions are posted on the QNRF website.
May 1–June 22, 2007	QNRF responds to questions and inquiries from applicants submitted online and posts responses on its website.
June 8, 2007	Detailed proposal submission instructions are available online at www.qnrf.org.
July 1, 2007	Letters of intent are due.
August 1, 2007	Proposals are due.
January 2008	Awards are announced.

accordance with its guiding principles. The communication policy it adopted ensured that all applicants had access to the same information for their proposals.

Immediately after the release of the RFP, QNRF began extensive outreach to publicize the program and generate interest. QNRF staff made a number of public presentations at which they fielded questions from prospective applicants. The questions were important, since some required decisions about policy issues the QNRF-RQPI team had not yet considered. A QNRF-RQPI committee developed answers to the policy questions and posted responses publicly on the QNRF website. QNRF also took questions online and posted answers on the website. This process was important for continuous policy development, and it enabled the first cycle to be treated as a learning cycle. It also helped to identify parts of the RFP that might need further clarification for future rounds.

After June 2007, QNRF no longer communicated directly with applicants about their questions. Following open communication with silence was in line with the practices of NSF and NIH to ensure equal access to information. Limiting communication after a certain date would also help manage the workload of the small QNRF staff. QNRF continued the practice of silence until just before it announced the awards.

Developing an Online Application Process

QNRF believed that it was important for the NPRP to have an online application process, for several reasons. First, it was anticipated that the number of applications—possibly several hundred—would necessitate such an approach, since manually processing them with multiple peer reviews each would be very difficult and time consuming. Second, it could be very easy to make errors when doing the processing manually. It was concluded that greater standardization of information processing and sharing could resolve the problem, so a web-based solution was implemented, one that supported a basic level of web functionality.

Setting up an online system in a very short time would be challenging and would also entail risk. The NSF's development effort had taken 11 years from original concept to final implementation. It was far from certain that QNRF could create an online system quickly enough to be ready to process the first round of NPRP applications. Nevertheless, web-based systems are common today, and a new website would have a fairly good chance of meeting the program's needs. In addition, successfully managing the administrative processes of the first cycle of NPRP was important for QNRF's credibility.

The QNRF-RQPI team understood the need to protect sensitive financial information about research institutions, salary information about researchers, and IP in the proposals. Therefore, the team consulted with Q-CERT[1] to review the security of the web application. Q-CERT reviewed the application architecture and provided recommendations for improving the overall security of the website, as well as protecting the privacy of the data and the confidentiality of the website's users.

Because there is always a risk of problems with a new system, QNRF required applicants to submit paper proposals in addition to the online submissions. This proved to be a prudent decision. While the website worked for many proposals, it presented significant user-interface

1 Q-CERT was established in Qatar in cooperation with the original CERT organization at Carnegie Mellon University to serve as the national cyber-security program for the State of Qatar.

challenges, leading some applicants to fail to upload all the parts of their proposals. Several online applications were missing sections that appeared in the paper applications. This dual-submission approach to the application process ensured that QNRF had both the functionality of a web-based system and a backup that hedged against errors.[2]

In addition to developing an online application system, QNRF developed an online peer-review system; that system was also reviewed by Q-CERT, since reviewer confidentiality was important. More than 700 peer reviews were filed online in the first cycle.

After the submission process, QNRF received significant feedback from applicants and peer reviewers about challenges and ways to improve its online system. Still treating the first cycle as a learning cycle, QNRF welcomed constructive comments. (In later cycles, the UREP used the NPRP online system, which QNRF believed would standardize and improve the UREP proposal process as well.)

Initial Response to the NPRP's Request for Proposals

Following an administrative screening process described in the RFP and with RQPI's help, QNRF staff screened the 200 proposals that had been submitted before sending them to peer review. Those that met the administrative criteria were sent to review, and those missing sections or significantly violating application rules were declined. The QNRF-RQPI team was well aware that the first NPRP award cycle would constitute a learning process for both QNRF staff and many of the researchers. Consequently, QNRF chose to be somewhat lenient with regard to the administrative criteria. The QNRF-RQPI team also acknowledged that failure to follow administrative instructions could mean that the instructions in the initial RFP lacked clarity. However, almost 90 percent of the proposals passed the first stage of administrative review and went on to peer review.

Conducting the NPRP's First Round of Peer Review

The Process and the Challenge

Finding peer reviewers for 200 proposals, matching them to proposals, collecting scores, and making grant decisions based on those scores was not easy. The requirement in the RFP that applicants file a letter of intent had been designed to enable QNRF to begin finding peer reviewers early in the process. However, the results were cause for concern: QNRF received more than 300 letters of intent, and although not all of those who submitted letters of intent submitted proposals, QNRF staff had to seek peer reviewers for all of them.

Believing that a number of peer reviewers per proposal would make scoring more robust, QNRF set a target of five reviews for each submission. While the unexpectedly large response was certainly good news, QNRF was going to have to line up almost 1,000 peer reviewers in time to meet the schedule set for the announcement of awards. Funding decisions were supposed to be issued by the end of November, less than four months away.

---

[2] The QNRF team cross-checked the online submissions with the paper submissions. When the submissions differed, the team sent the version that appeared to be the one intended to the reviewers.

All of the bins named in the RFP contained some proposals. The industry and engineering bin garnered the most, followed by computer science and IT and health and biosciences (see Figure 6.1).

The QNRF-RQPI team shared the work of finding appropriate peer reviewers.[3] The criteria for reviewers generally required that they

- Be at least associate professors or the equivalent
- Have authored at least 20 peer-reviewed publications, some of which had to be recent
- Be employed at a research institution.

These requirements are more stringent than those of many academic journals. At some journals, graduate students or postdoctoral researchers may be used as referees, provided they are sufficiently knowledgeable. Although the very high NPRP standards did make the task of finding reviewers more challenging, they provided clear and transparent criteria that increased the confidence of the institutions whose proposals were being reviewed.

By mid-November 2007, nearly all of the first-round proposals had been assigned peer reviewers. QNRF actually went quite far toward meeting its original target of five reviewers per proposal: Each proposal had from three to six reviewers, and only 33 had as few as three reviewers.

Figure 6.1
Distribution of Peer-Reviewed Proposals

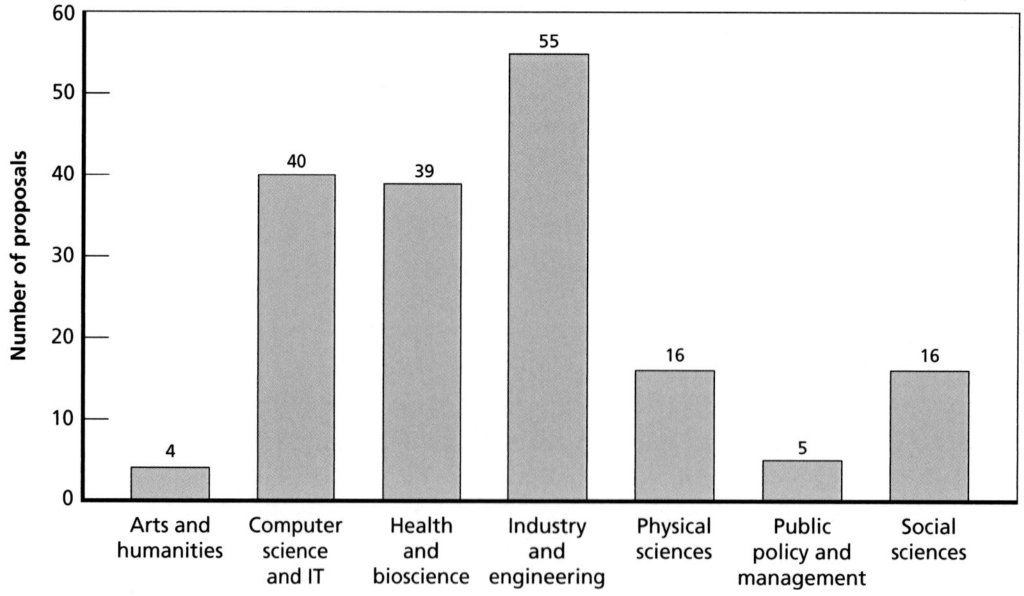

RAND *TR722-6.1*

[3] The Civilian Research Development Fund (CRDF) assisted in this effort. CRDF is a nonprofit institution based near Washington, D.C., that funds scientists from countries in turmoil, such as the former Soviet Union in the 1990s. It has recently turned its attention to scientists in the Middle East. CRDF maintains a database of 18,000 experts in science and engineering.

Lessons Learned from the First Round of Peer Review

The first cycle of peer review provided valuable experience about what worked and what did not. Using the letters of intent to enable program officers to get a head start on the search for reviewers was not as helpful as had been hoped. Many of the letters did not contain enough information to identify qualified reviewers for the proposals that followed; in other cases, applicants had altered their research ideas by the time they submitted proposals. The peer reviewers who were identified on the basis of the letters of intent were often not appropriate for the actual proposals submitted.

QNRF and RQPI used several approaches to recruit peer reviewers. They identified papers published in specific research fields and contacted the authors, but the response rate was relatively low. The QNRF-RQPI team attributed this, in part, to QNRF being unknown and to having started the search during the summer, when many academics are on vacation and not responding to correspondence.

Personalizing efforts to recruit peer reviewers, however, worked well. QNRF and RQPI staff wrote letters of introduction to possible reviewers and often followed up with phone calls. Although this was time consuming, it was worth the effort, because candidates were more likely to agree to review proposals, and contacts built up in this way would make each future cycle easier.

QNRF began to build a database of peer reviewers, so that finding those for subsequent cycles would be less onerous than the search in the first cycle. It also decided to include a search tool in its database that enabled résumés to be searched by keywords. This allowed QNRF staff to rely on the computer, rather than having to read through all the résumés.

Making NPRP Award Decisions

Once the peer reviews had been received, the QNRF leaders had to make the award decisions. There were three main issues to be resolved: Should the decisions be based on scores only or on panel decisions? What summary score should be used? And how should money be allocated among the bins?

QNRF decided to base funding solely on the numerical scores from the peer reviewers, for two reasons. First, it minimized the burden on an already overstretched QNRF staff that would be unable to reorder or manage panels of peer reviewers within the time lines set for decisionmaking. Second, QNRF preferred to maintain defensible policies and transparency of decisionmaking for the research community; numerical scores were clear, understandable, and sent a message of a rigorous, merit-based process.

Then it was necessary to decide what summary measure should be used to make decisions. The measures considered included the simple arithmetic mean of each of the total scores and the sum of the medians of the seven subscores. While the simple arithmetic mean is arguably more sensitive in that it intrinsically depends on the ratings of each of the reviewers on each of the components of the total score, the sensitivity can lead to problems. The mean is not a robust measure with respect to erroneous or outlier data. The sum-of-medians approach, in which the summary measure is calculated by summing the "middle score" for each of the component scores, is less sensitive.

Figure 6.2 plots the mean score and the sum-of-medians score for each proposal. The points that fall below the diagonal are those in which the consistently low rating of one reviewer

Figure 6.2
Mean and Sum-of-Median Scores of the Proposals

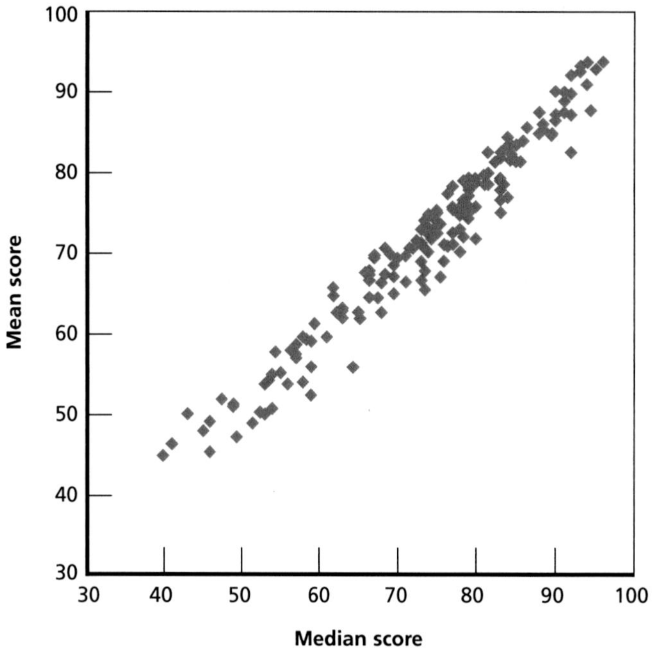

RAND *TR722-6.2*

on one or more of the component scores had an appreciable impact on the final score when the mean was used. This finding led QNRF to use the sum-of-medians approach, which would be more robust to outliers in the data or to reviewers who consistently rated higher or lower than others. This approach gave the proposers the benefit of the doubt.

The final substantive decision concerned the allocation of funding among the bins. Should funding be based on an absolute, across-the-board score for all of the bins or a relative score (a percentile approach) in which proposals would be evaluated against competing proposals in the same bin? Both approaches are used by reputable funding agencies (Johnson, 2008). The absolute approach has the advantage of holding all proposals to a uniform standard, which gives no special advantage to those in a particular area of research. The relative-standard approach is advantageous in situations where scoring norms differ appreciably across research fields.

While many proposals would be funded under either criterion, the choice of standard does have some effect at the margin. Figure 6.3 shows the relative distribution of grants across fields of study and across major institution groups.

The various research fields had fairly even scores and percentages of success, so it was not necessary to allocate money according to bin or to determine different cut-off scores for each bin. The absolute criterion slightly favored industry and engineering, public policy and management sciences, and physical sciences and the environment. The percentile criterion slightly favored computer science and IT, health and bioscience, and social sciences. However, the differences were not great enough and the sample size was not large enough to determine whether peer reviewers in certain fields routinely scored higher than those in other fields.

Figure 6.3
Distributional Implications of Using Absolute Versus Percentile Scores to Fund Proposals

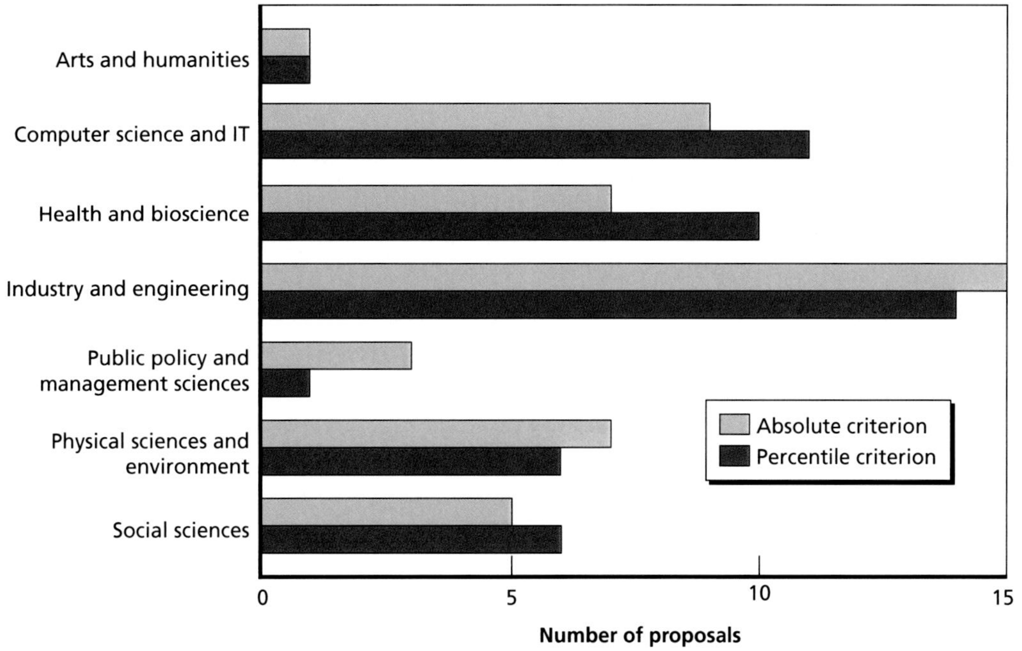

RAND *TR722-6.3*

After reviewing the two alternatives, and upon the recommendation of the start-up manager of QNRF, the QF Board of Directors chose the absolute standard for the first cycle of the NPRP.

Awarding Grants

In the first NPRP cycle, QNRF funded 47 projects in amounts ranging from \$20,000 to \$250,000, for a total of \$25 million over three years. QF would decide on budgets for awarded proposals for future years on the basis of how the program was faring.

Once the awards had been announced, QNRF negotiated a memorandum of understanding (MOU) with each of the receiving institutions. The MOUs addressed IP issues, financial agreements, administrative procedures, ethical standards, and other factors. RQPI staff created a draft MOU, which was revised by QNRF with the assistance of QF Finance. Using this draft MOU, QNRF negotiated agreements with each of the institutions involved. One challenge in securing agreements was that many recipients had their own research and IP rules. While parts of the MOUs were standardized, others had to be tailored to individual circumstances. For example, the universities in Education City have to follow national or state laws of their home campuses.

In early 2008, QNRF transmitted funds to the grant recipients. In only about a year and a half, QNRF had established itself administratively and had launched initial rounds of funding for both of its key programs. Qatar now had in place the basic research-granting infrastructure to support its goal of becoming a world-class center of academic and research excellence.

Lessons Learned in Designing and Launching the NPRP

The QNRF-RQPI team learned some valuable lessons during the first cycle of the NPRP. Most significantly, it learned that the program could be considerably improved. One cycle did not enable definitive conclusions to be drawn about how well the overall NPRP structure was meeting the QNRF goals. This suggested that a thorough evaluation should be made after several cycles had been completed. Areas for this evaluation might include policies and administration, the point system, the bins, budgeting, the peer-review process, and peer-reviewer selection. After more experience with administering the programs, such an evaluation could help to strengthen the organization and could assist in building the research culture and infrastructure in Qatar.

That said, the QNRF-RQPI team had already found several ways to improve the administration of the program for the next cycle.

First, QF decided to increase the maximum grant amount to $350,000 per year. Given the costs of conducting research in Qatar, which are higher than those in the United States or Europe, the QF leadership believed that the increased maximum amount would better serve the needs of the research institutions in Qatar.

Second, outreach to the research community in Qatar and abroad had been very important in creating awareness of the program, so QNRF decided to do even more outreach, both inside and outside of Qatar, in future cycles to give potential applicants the information necessary to prepare proposals. Part of this outreach effort consisted of strengthening QNRF's website to make it more user-friendly and to make key information more easily accessible.

Third, feedback from the research community in Qatar provided valuable information on ways to improve the program. Applicants and grant recipients were well situated to offer suggestions on how to make the program more effective.

Fourth, the process in the first cycle demonstrated the need to have award administration concepts and policies, as well as a team, in place for award administration to deal with projects after grants were awarded. QNRF needed to develop a funding-agency culture.

Fifth, the IP policy chosen was a useful starting point, but it needed to be further developed. After the NPRP grants were awarded, QNRF completed individual IP agreements with all of the grant recipients in Qatar. However, negotiating the actual agreements took a significant amount of effort. Many institutions had complicated considerations and preexisting constraints that they brought to the negotiating table. After several years of experience with its IP system, QNRF might consider performing an evaluation of the strengths and weaknesses of this approach and adjusting it accordingly. It might also aim for an appropriate level of consistency and flexibility in its agreements with partner institutions. Regardless of the solutions chosen, it is important to maintain an appropriate system of incentives that encourage researchers to create, develop, and exploit their IP.

Finally, the experience of the first cycle showed how important a well-functioning online proposal-submission process was. While the QNRF's online submission system met basic needs, it needed to be made more user-friendly. QNRF used feedback from users to restructure the online submission process so that it would better meet the needs of the program in the next round. As website development is not one of its core competencies, QNRF partnered with another organization to accomplish this objective.

CHAPTER SEVEN

Looking Ahead: Next Steps and Future Challenges

In about a year and a half—from July 2006 to January 2008—QNRF grew from a concept on paper to a research-granting organization that awarded more than $30 million to teams of researchers both inside and outside of Qatar. By early 2012, it had awarded about $500 million and had developed the beginning of a research-granting infrastructure to support a research community in Qatar. This is a solid start toward QF's vision of QNRF using "research as a catalyst for expanding and diversifying the country's economy; enhancing the education of its citizens and the training of its workforce; and fostering improvements in the health, well-being, environment and security of its own people and those of the region" so that Qatar can "distinguish itself within the region and world as a cosmopolitan nation that embraces scholarly excellence, innovation, creativity, inclusiveness, and merit" (Greenfield et al., 2008).

Elements necessary for continued development seem to be in place. The most important component—the research projects awarded during the first NPRP and UREP cycles—is already there. However, some difficult work lies ahead for QNRF to truly establish itself as a research-granting body of international caliber.

This chapter summarizes QNRF's principal achievements and implications for further analysis and identifies opportunities that QNRF could take advantage of in the future.

Progress from 2006 to 2008

Between 2006 and 2008, QNRF established the first international research-granting organization in the Middle East. Its accomplishments included

- Design of guiding principles for its research programs and policies based on its vision and goals
- Creation of a matrix management structure in which a small staff could carry out the functions necessary for start-up
- Establishment of an interim governance structure in which a steering committee consisting of local members could meet regularly and make decisions quickly
- Implementation of the UREP, awarding approximately $5 million to faculty and undergraduate research teams in its first four funding cycles
- Design and implementation of the NPRP to support professional research in Qatar
- Formation of an IP policy to accommodate the many different IP systems of grant recipients inside and outside of Qatar

- Development of administrative policies, an online proposal-submission system, and a peer-review process
- Awarding of about $25 million in NPRP grants to researchers in the Education City branch campuses, Qatar University, Hamad Hospital, and other institutions in Qatar, in collaboration with research institutions in about 30 countries.

Possible Future Changes

The government of Qatar made a commitment in 2007 to spend 2.8 percent of its revenue on research and development (R&D). It subsequently issued Decree-Law No. 24 of 2008, which names QNRF as an entity that will implement and administer some of the new R&D funding and projects. In 2012, QF began a process to involve stakeholders in determining future priorities for investment in research. QNRF's potential new responsibilities as a funder argue for a careful examination of its focus for future programs and its governance structure. Some suggestions that QNRF could pursue to ensure that it can meet its current vision as well as potential new demands are presented below.

Create Focused Research Priorities and Additional Programs

As Qatar invests more of its revenue in research, its leadership may choose to mix two approaches to achieving Qatar's national research strategy: (1) focusing funding opportunities in areas that appear to be of greatest importance to Qatar and (2) enhancing QNRF's complex adaptive approach, in which researchers propose areas they think are important, allowing new ideas to "bubble up." As Qatar's research leaders determine research priorities, QNRF may examine how it can best support those aims and encourage creativity in researchers in Qatar.

Conduct an Evaluation of Current Policies and Programs Every Few Years

QNRF, its programs, and its policies were based on international best practices in research-grant management, tailored to Qatar's unique circumstances. The original structures proved sufficient to meet initial needs—i.e., they were adequate to get the organization started. However, addressing the needs of Qatar's research community and the nation's research goals will require changes in programs and policies over time to continuously improve effectiveness. Those changes should be based on analyses and self-evaluations of how QNRF can better meet its objectives, as well as on feedback from researchers, research organizations, peer reviewers, and other stakeholders. QNRF has already begun the process of self-evaluation by soliciting feedback in the early cycles of the UREP and the first cycle of the NPRP and finding areas for improvement after each funding cycle. In addition, QNRF should undertake a thorough evaluation of its programs and policies every few years. The evaluation should include the appropriateness and effectiveness of

- The UREP
- The NPRP (including its approach, bins, budgeting, point system, and peer-review process)
- Grant management strategies, policies, and structures
- IP policies
- The potential need for additional programs.

Strengthen Long-Term Governance Structures

Management and governance will need to evolve as QRNF grows and expands its responsibilities. At QNRF's start-up, an interim steering committee was put in place. The interim steering committee was able to start working immediately, meet regularly, and make decisions quickly, and it included stakeholders who could shape decisions to meet the needs of the research community. Now that QNRF is up and running, the establishment of a strong, experienced governing board is very important. The governing board will need to establish QNRF's overall direction, set policy, and provide institutional oversight. It will also need to ensure that QNRF's mission and goals are implemented in a sound and timely manner. QF asked RQPI to provide a plan for a governing board; this plan is described in an RQPI report.[1]

Define and Establish Measurements for a Vibrant and Innovative Research Culture

Discussions throughout the creation of QNRF concerned the need to create a "research culture" in Qatar. However, questions about what a research culture is and how it can be measured arose repeatedly. While many used the phrase and seemed to have an intuitive sense of what it meant, the concept was never fully defined. QNRF asked RQPI to undertake further research to define research culture and to suggest ways to assess progress toward it. This analysis of research culture is also discussed in Cecchine at al. (2012). This would also be a significant contribution to the development of important concepts of research management. In addition, QNRF might develop a robust set of measurements for its research outcomes and success.

---

[1] This document (Cecchine et al., 2012) discusses fostering a research culture and sustaining QNRF well into the future.

APPENDIX A

Undergraduate Research Experience Program Request for Proposals

Qatar Foundation
Qatar National Research Fund

Undergraduate Research Experience Program
Request for Proposals
2006

1. Overview of the Program

The mission of the Qatar National Research Fund (QNRF) is to advance knowledge and education by supporting original, competitively selected research in: the physical, life, and social sciences; engineering and technology; the arts; and the humanities. It will provide opportunities for researchers at all levels, from students to professionals, in the private, public, and academic sectors. The QNRF strives to encourage and support high-quality basic and applied research serving Qatar's national needs and interests.

Recognizing the importance of introducing research in undergraduate education in Qatar, the QNRF is initiating the Undergraduate Research Experience Program (UREP). This program aims to engage undergraduates under the mentorship of faculty members in all universities in Qatar on research projects related to Qatar's national needs.

The UREP will promote "Learning by Doing" and "Hands-On" research activities as effective methods for undergraduate education. Students will gain experience in research collaboration with faculty, postdoctoral fellows, graduate students and other undergraduates or research staff in Qatar.

2. Benefits of UREP

The following are some of the benefits to parties involved in the UREP:

(i) Undergraduates
- Gain experience in problem-solving
- Develop communication skills and work independently
- Understand research methods, ethics, and rules of conduct
- Understand the link between academics and other careers
- Work with a diverse group of people and in varying environments
- Network with faculty beyond the classroom
- Obtain knowledge on conducting research that could lead to graduate studies
- Gain credit hours; as per the bylaws of each educational institution

(ii) Faculty
- Gain student support and labor for their research
- Network with students beyond the classroom
- Identify potential students for postgraduate study and research
- Network with industry

(iii) Qatar

- Develop student capabilities
- Further the research culture
- Expand knowledge in Qatar and about Qatar

3. Solicited Proposals

The UREP is intended to maximize faculty-student interaction and enhance the learning experiences of undergraduate students, by encouraging faculty to integrate students into their ongoing scholarly and professional activities. In consideration of the academic diversity of Education City and Qatar University, the UREP seeks to support a broad variety of projects. These may include basic and applied research and projects in the arts, sciences, engineering, and humanities. All UREP projects must have substantive educational content.

Proposals may be for projects where students work individually with specific faculty, or they may be for projects where teams of students work either with individual faculty or faculty teams. The UREP will also consider collaborative projects involving more than one university in Qatar. Selection of undergraduate participants and research topics, and detailed management and reporting will be the responsibility of the lead faculty member receiving the award.

Proposals should be prepared by faculty who will lead the research involving undergraduates. The Appendix: "Proposal for Undergraduate Research Experience Program" is the suggested format of the proposal. Ideally, students should be involved in the proposal process, to the extent possible, as a learning experience.

As part of their research experience, students are expected to receive training in research methodologies and ethics, information retrieval and sharing, and scholarly communication.

Each project should culminate with an appropriate report of the completed work (for example, a formal research paper, design project or similar demonstration of scholarly/creative accomplishments).

4. Funding

The UREP grants are primarily intended to support students and mentorship activities by the faculty. In creating budgets, emphasis should be placed on support for student and faculty time and effort.

A total budget per student up to $10,000 may include:

- A $4000 student stipend.
- A $2000 faculty stipend for mentorship activities.
- (Optional) Up to an additional $4000 for:
  - purchasing and maintaining supplies and equipment;
  - travel for research and participation in conferences; and/or
  - publishing.
  - (Reimbursement for expenses listed and approved in the original budget of the proposal will be upon submission of receipts. Funding cannot be used for institutional overhead.)

This grant does not preclude acceptance of grants from other sources, if the other grants do not duplicate coverage of the same expenses.

5. Eligible Participants

The Undergraduate Research Experience Program shall be open to all interested faculty and undergraduates studying at Qatar's universities in all disciplines.

6. Deliverables

The following deliverables are required by QNRF.

(i) Progress Report

A one-page progress report will be submitted 4 months after receipt of funding. It should highlight progress made, results obtained, and deviations from the original objectives or research plan, if any.

(ii) Final Report

Each project will culminate with an appropriate project or report of the completed work: for example, a formal research paper, design project or similar demonstration of scholarly/creative accomplishments.

*(iii) UREP Conference (Participation Optional)***:**

Students will have the opportunity to present their research to other undergraduates and faculty participating in the program in an interdisciplinary UREP Conference hosted by QNRF. Awards and recognitions will be given to distinguished students and projects at the conference.

7. Proposal Evaluation Process

QNRF will seek input from at least 3 independent reviewers for each proposal. The reviewers will score each proposal against the merit review criteria shown below. A summary narrative and accompanying rating will be provided by each reviewer; these reviews will be treated confidentially. Copies of the reviews, excluding the names of the reviewers, will be forwarded to each proposer with an explanation of the decision to award or decline funding. QNRF will base its funding decisions heavily on these reviews and a consideration of how to allocate available funds to best support the QNRF mission.

Evaluation Criteria

(30 points) *Anticipated benefit to the undergraduate student.* This criterion addresses how well the proposed activity advances discovery and understanding while promoting teaching, training, and learning. It includes an assessment of the nature of tasks to be performed by the student, number of hours for the student, the anticipated final student product, and the anticipated level of faculty-student interaction.

(20 points) *Anticipated benefit and relevance of the proposed activity to Qatar.* This criterion addresses how the project can further knowledge of the physical, life, and social sciences; engineering and technology; the arts; and the humanities in Qatar or about Qatar.

(20 points) *Intellectual merit of the proposed activity.* This criterion addresses how important the proposed activity is to advancing knowledge and understanding in the field. It includes the soundness, originality, and creativity of the research or project design and anticipated products. It describes the nature of the final research product and the student product to be completed during the research experience (for example, a report, poster session, oral presentation, creative product, or thesis proposal).

(30 points) *Mentoring and oversight.* This criterion addresses plans for the oversight, supervision, and mentoring roles by the proposing faculty, including plans that will ensure the development of student-faculty interaction and plans for program evaluation. It describes the process and criteria for selecting student researchers, to assure that students who participate have demonstrated academic competence, are well suited to the project structure and content, and will benefit academically from the experience

8. Timeline

The following dates represent the major milestones of the UREP:

October, 2006:	Solicit proposals, publicize UREP in public Question and Answer sessions.
December 15, 2006:	Deadline for receipt of 1st cycle of proposals.
February, 2007:	Announcement of awards.
March, 2007:	Signing of Memoranda of Understanding (MoUs) and availability of funds for winning proposals.

Appendix: Proposal for Undergraduate Research Experience Program

A. Title of Research Program

B. Participant Information

Faculty	**Details**
Title and Position	
First Name	
Last Name	
University	
School/Department	
Postal Address	
Phone	
Fax	
Email	

Undergraduate(s) (attach additional forms for each additional student)	**Details**
First Name	
Last Name	
University	
School/Department	
Postal Address	
Phone	
Fax	
Email	

C. Abstract of the Research Project in English (and Arabic)

D. Role of Undergraduate Researchers (include number of hours per week per each participant and key tasks to be performed)

E. Summary of Expected Benefits to the Student(s), Institution, Qatar, and/or Community

F. Detailed Budget of Expected Expenses

G. Timeline/Milestones (include description of the final research product, and estimated time of completion, to be produced by the student(s))

H. Evaluation Plan

A plan for the evaluation of the proposed project, to measure the success of the project in achieving its goals, particularly the degree to which students have learned from the experience. The evaluation plan is an important part of the proposal, but proposers have much latitude in designing a plan that best suits their particular project.

I. Proposal Evaluation Criteria

The proposal should explicitly address how this project will fulfill the evaluation criteria outlined in Section 7: Proposal Evaluation Process above.

J. Language of the Proposals

Proposals may be submitted in either English or Arabic. If proposals are submitted in Arabic, then the personal information form on p. 6 of this application AND the Abstract of the Research Project (item C above) must also be translated into English.

APPENDIX B

National Priorities Research Program Request for Proposals

مؤسسة قطر
Qatar Foundation

الصندوق القطري لرعاية البحث العلمي
Qatar National Research Fund

NATIONAL PRIORITIES RESEARCH PROGRAM

Request for Proposals

April, 2007

Table of Contents

Appendices

1. Overview of the National Priorities Research Program

The mission of the Qatar National Research Fund (QNRF) is to advance knowledge and education by supporting original, competitively selected research. It will provide opportunities for researchers at all levels, from students to professionals, in the private, public, and academic sectors. For its flagship program, the National Priorities Research Program (NPRP), QNRF presents a Request for Proposals (RFP). Proposals are due on August 1, and QNRF will award its first grant in November 2007. The National Priorities Research Program is the largest grant funding activity of the QNRF and the primary means by which the QNRF will seek to address key national, regional, and global needs through research and to pursue research opportunities for which Qatar may have a comparative advantage.

2. Strategic Goals and Research Portfolio of QNRF

QNRF seeks to support a broad variety of projects. These may include basic and applied research projects in: Industry and Engineering; Physical Sciences; Arts and Humanities; Social Sciences; Public Policy and Management Science; Computer Science and Information Technology; and Health and Biosciences. Proposals are invited in all of these disciplines.

Proposals should address Qatar's national interest and QNRF's mission by proposing research that will lead to:

1) accumulation of valuable human capital or a sustainable research capability in Qatar
2) answers to research questions of vital interest to Qatar
3) recognition of Qatar due to the regional or global significance of the proposed research.

Appendix 1 includes a list of example research areas relevant to Qatar's national needs for this first cycle of the NPRP. Proposals for other research projects that applicants see as important to Qatar will be welcomed and encouraged.

3. Benefits of the NPRP

- **Service to Qatar**. The NPRP will serve the people of Qatar, as well as its business, academic, and government sectors. It will strive to build bridges between businesses, government, academia, and others in Qatar, and between Qatar and the international research community.
- **Collaboration and Sustainability**. The NPRP will fund programs that encourage collaboration among researchers in Qatar, as well as between researchers in Qatar and their international colleagues by putting special emphasis on multi-investigator, multi-year research projects.
- **Identifiable National Benefits**. As noted above, all work supported by the NPRP will have the potential to yield identifiable benefits to Qatar. Benefits to Qatar may include development of human capital in Qatar, answers to research questions of direct interest to Qatar, or international recognition of Qatar due to the regional or global significance of the proposed research.
- **Active Publication, Promotion, and Outreach**. The QNRF will publish, archive, and summarize research results to promote Qatar's research activities, both locally, regionally, and internationally. QNRF will facilitate interactions within Qatar's research community and reach out to the wider community of Qatar.
- **Commercialization and Diversification of the Economy**. The NPRP will support both basic and applied research. Some of this can lead to expertise in new sectors and a diverse, thriving economy.

4. Project Design and Program Eligibility

QNRF encourages multi-year, multi-disciplinary, and multi-institutional projects, as well as proposals for small individual projects. Teams of researchers may come from a single institution, or involve multiple institutions, and they may span academe and industry. QNRF encourages collaborative projects partnering researchers on different campuses or partnering academia and organizations outside academia. QNRF also encourages partnerships between institutions in Qatar and outside of Qatar.

Proposals are welcome for projects spanning between one and three years. For projects longer than one year, funding will be renewable on a yearly basis following QNRF review of a progress report.

Grants will be made to institutions and organizations only; research may be under the auspices of academic institutions, businesses, or governmental or non-governmental organizations.

As a general rule, a significant portion of research funded by the QNRF must be conducted in Qatar. However, it is fully expected that a coordinated research program funded by QNRF could take place in part at other venues. Principal Investigators may collaborate with accredited research organizations in other parts of the world. Principal Investigators in Qatar need not be citizens of Qatar, but at least one Principal Investigator or co-Principal Investigator per project must be a resident of Qatar.

5. Letter of Intent

QNRF requests a letter of intent from applicants by July 1, to be submitted online. This document is to provide QNRF with the estimated number of applicants and their specific areas of research for the purpose of planning the peer review process. Information included in the letter of intent is considered draft information for planning purposes and Principal Investigators may modify information in the final proposal.

The Letter of Intent should provide the following:

- Name, institution and contact information of the Principal Investigator
- Research field that will be addressed in the proposal:
    - Industry and Engineering (energy, aviation, environmental impacts, etc.)
    - Physical Sciences (physics, chemistry, ecology, etc.)
    - Arts and Humanities (design, Islamic studies, culture, ethics, etc.)
    - Social Sciences (economics, education, regional history, law, political science, etc.)
    - Public Policy and Management Science
    - Computer Science and Information Technology
    - Health and Bio-sciences
- Title and brief description of the proposal (Do not exceed the space provided)
- Time period required to complete the research project (e.g. 1, 2 or 3 years)
- Approximate budget total

6. Proposal Preparation

Prior to receiving a proposal, QNRF requests a letter of intent one month in advance, for planning the Peer Review Process.

Please read instructions carefully before preparation of the proposal (see online forms for instruction). See "Submit Online" at www.qnrf.org for more detailed instructions and guidelines on how to prepare your proposal as well as for the forms for the application itself. Online forms and detailed instructions for the items below will be available on June 1.

Proposals must include:

- Face page (or cover page) signed by the Principal Investigator containing the title of the project; the Principal Investigator's name, degree, position title, institutional mailing address, and contact information; dates of proposed period of support; and the budget requested.
- A letter signed by the Principal Investigator's dean or equivalent to QNRF in support of the proposal and the Principal Investigator, verifying that the proposal complies with the institution's policies and certain QNRF policies as stated in this RFP.
- A letter signed by the Principal Investigator's institution financial officer certifying that the proposal complies with the institution's research budget policies and with certain QNRF policies as stated in this RFP.

- Description page, performance site(s), and names of Principal Investigator(s).
- Description of the benefit of the research project to Qatar, the region, or the world community.
- CVs or biographical sketches of the Key Investigators
- Resources available to perform the research proposal.
- Research Plan (25 page limit)
    - Specific Aims and Timeline (one page recommended)
    - Background and Literature Survey
    - Significance
    - Preliminary Data or Studies
    - Research Design and Methods
    - Evaluation Plan
- References (Literature cited)
- Compliance with national ethical rules and regulations followed in Qatar and any other countries where the research will take place.
- Letters of support from consultants, collaborators, institutions, etc. (if applicable)
- Detailed Budget for the initial year (year 1).
- Budget for entire proposed period (2 or 3 years), if the proposal is greater than one year
- Budget Justification, according to the instructions
- Previous Publications (If available, attach PDF files of not more than 5 publications by the applicant researchers that are pertinent to the proposed project.)
- Checklist

Proposals must be submitted in English, except in a limited number of fields in which the primary international language of research is Arabic (Islamic studies and Qatari law). If Principal Investigators in a different field propose to submit an application in Arabic, they must obtain QNRF approval before June 15. In such a case, an abstract in English must be submitted.

7. The Submitting Institution's Role in Proposal Preparation

The submitting institution bears final responsibility for ensuring that all proposals coming from individual researchers within it comply with:

- QNRF policies as stated within this RFP
- The institution's internal policies
- Applicable laws and regulations where the research is being conducted

The Dean or equivalent at the submitting institution must verify that the proposals comply with QNRF and institutional policies, and the Dean must submit a letter of endorsement certifying that he or she has done so. In addition, the Chief Financial Officer or equivalent at the submitting institution must also thoroughly check proposals to ensure that the budgets comply with QNRF guidelines and the submitting institution's policies, and submit a letter of endorsement to that effect. Forms for these letters are provided among the forms for the application. (See www.qnrf.org.)

For policy questions for which the QNRF has not defined guidelines, the submitting institutions must ensure that proposals follow the institution's policies.

Proposals submitted without these two letters of institutional endorsement will be declined.

For joint proposals, the lead institution(s) must provide these letters.

8. Eligible Expenses and Funding

QNRF is pleased to announce that it plans to fund a total of $10 million in research during the first NPRP cycle in 2007-2008.

For the NPRP's first cycle, QNRF will fund grants ranging from $20,000 up to $250,000 per proposal per year. This will ensure that various projects with a broad range of research areas are funded by QNRF.

QNRF after due consideration of well reasoned requests for a higher level of funding may invite proposals for projects of exceptional merit and extraordinary national importance. Uninvited proposals for a higher level of funding will be declined.

At least one Principal Investigator per project and a significant portion of research staff must reside in Qatar during the project. QNRF encourages projects involving teams of more than one institution, cross-campus collaboration in Qatar, academic and non-academic organizations, and organizations both inside and outside of Qatar.

For projects longer than one year, funding will be available annually upon QNRF review of a satisfactory progress report based on an evaluation plan laid out by the team itself in the proposal. A satisfactory progress report demonstrates that the research teams have met the milestones they set for themselves in their proposal.

As noted in Section 7, budgets must follow the instructions included in the online budget instructions and must comply with the approved budget policies for research set by the grantee institutions. Budgets must be accompanied by a letter of endorsement from the management of the Principal Investigator's finance department, certifying that the budget is in accordance with the institution's budget policies on research grants.

Research expenses that may be funded include:

Compensation of research staff:
Research staff includes the following: Principal Investigators; other senior or junior researchers; post-doctoral fellows; graduate students; research assistants; technicians; and programmers.

- The project can charge toward the compensation of research staff according to percentage of time spent on the project. For example, if a researcher will spend

20% of his or her time on the project, the project can charge for 20% of his or her compensation package.

- Researchers may use the grant money to buy their time from other duties such as teaching or committee work, as per the policies of their institution.
- For those researchers who come to Qatar specifically for work on an NPRP research project, a percentage of accommodation, transportation, and graduate student tuition, equal to the researcher's percentage of time spent on the project, may be charged to the project. This is not allowed for researchers already in Qatar.
- Over 50% of the total research hours must occur in Qatar.

Grantee partners outside of Qatar:

- In cases in which two Principal Investigators work at different institutions (either both inside of Qatar, or one inside of Qatar and one outside of Qatar), funding will be given directly to the separate institutions involved.
- At least 65% of the funding must be under the administration of an institution in Qatar for use of the researchers residing in Qatar.

Research equipment and material purchases, repair and maintenance:

- Proposals must include documentation of the prices of the items requested. For example, a proposal could include a printout from an internet order or a photocopy of a catalogue.
- Equipment during the project will remain property of the QNRF. Ownership of equipment after the duration of the project will be finalized in the signed MOUs with the research institutions.

Research-related travel costs:

- QNRF will fund reasonable travel expenses within the QNRF's policies and the submitting institution's policies. Please see the budget instructions for detailed policies.
- Applicants must list estimated costs.
- If an award is granted, PIs must submit detailed trip reports in progress reports.

Miscellaneous Costs:

- Publication, communication, administrative translation, and other costs

Payment will be made to the grantee institutions according to the payment schedule below.

- **1 year projects:** 40% upon the signing of the agreement; 40% upon the six-month progress report; and 20% upon receipt and acceptance of the final report.
- **2 year programs:** 50% of the first year's funding will be paid upon signing of the agreement. The remaining 50% of the first year's funding will be paid upon receipt and acceptance of the six-month progress report. 50% of the second year's budget will be paid after receipt and acceptance of an annual progress report. 30% of that year's funding will be paid upon receipt and acceptance of the six-month progress report, with 20% of that year's funding paid upon receipt and acceptance of the final progress report.

- **3 year programs:** 50% of the first year's funding will be paid upon signing of the agreement. The remaining 50% of the first year's funding will be paid upon receipt and acceptance of the six-month progress report. 50% of the second year's budget will be paid after receipt and acceptance of an annual progress report, and 50% upon receipt and acceptance of the second year's six-month report. 50% of the third year's budget will be paid after receipt and acceptance of an annual progress report. 30% of that year's funding will be paid upon receipt and acceptance of the six-month progress report, with 20% of that year's funding paid upon receipt and acceptance of the final progress report.

9. Ethics, Safety and Regulations

Projects must comply with any relevant ethical considerations, safety standards, and local/national regulations where the research is conducted. For example, the submitting institution and Principal Investigator(s) are responsible to follow safety guidelines of the submitting institution or country related to chemical, biological, and radioactive materials and to follow ethical procedures for human subjects research.

In addition, please see Appendix 2 for QNRF's Code of Conduct for its research grants.

10. Proposal Evaluation Process

Reviewers of the proposals will be solicited from peers around the world with expertise in the substantive area of the proposed research. QNRF will seek input from five independent peer reviewers for each proposal. The overall review process for the 2007 cycle is comprised of the following steps.

Step 1: Initial Screening

In this initial step, QNRF staff will screen proposals against three criteria in a yes-no format, namely RFP Compliance, Cost Verification, and Benefit to Qatar. Proposals not in compliance will be automatically declined prior the review process.

RFP Compliance. Proposals must include all of the sections detailed in the request for proposals (see Section 5, "Proposal Preparation"), and they must not exceed the stated length restrictions. Proposals must also include verification from the Principal Investigator's Dean or equivalent that the proposal complies with QNRF policies as stated within the RFP as well as the institution's internal policies (see Section 7, "The Institution's Role in Proposal Preparation").

Cost Verification. Proposal budgets must be from $20,000 up to $250,000 per year. Section 8 of the RFP ("Eligible Expenses and Funding") details allowable expenses, including rules for compensation, equipment and material purchases, travel, publications, and institutional indirect rates. Proposals must contain a letter of endorsement from the Chief Financial Officer or equivalent of the institution, verifying that the proposal complies with university policies and QNRF guidelines as stated in this RFP. Proposals that do not adhere to these guidelines will not proceed in the review process.

Benefit to Qatar. As stated in QNRF's mission and described in the RFP, a primary goal for the NPRP is to fund research that has direct service to Qatar. These criteria are intended primarily to meet the NPRP primary goal of developing a research culture, human capital, and research infrastructure in Qatar. In order to pass this phase of the screening, proposals must meet the following conditions:

1. At least 65% of the proposed project funds must be administered by a university or other institution in Qatar and expended in or from Qatar
2. Over 50% of the research must be conducted in Qatar, as measured by person-days spent on the project
3. Either a Principal Investigator or co-Principal Investigator must reside in Qatar (but need not be a citizen)

Step 2: Categorizing Proposals Into Research Disciplines

Proposals will be grouped into research disciplines or "bins." Principal Investigators will select a discipline in which they want their proposals to be competitively reviewed. The disciplines are:

- Industry and Engineering (energy, aviation, environmental impacts, etc.)
- Physical Sciences (physics, chemistry, ecology, etc.)
- Arts and Humanities (design, Islamic studies, culture, ethics, etc.)
- Social Sciences (economics, education, regional history, law, political science, etc.)
- Public Policy and Management Science
- Computer Science and Information Technology
- Health and Bio-sciences

Step 3: Technical Peer Review and Proposal Ranking

Proposals that have passed the initial screening process (Step 1) will be reviewed within their discipline by five peer reviewers drawn from experts across the world. Peer reviewers will score proposals based on two sets of criteria: technical merit and research team qualifications, using the following questions as a guide:

<u>Technical Merit (75 points total)</u>

1. To what extent is the proposed research idea, approach, experimental design, data or outcome an original or new contribution to knowledge in the subject or to the literature? (Maximum 20 points)
2. How significant is the proposed activity to advancing knowledge and understanding within its discipline or across disciplines? How relevant are the projected results to problems in specific research area? Is the literature cited relevant and up to date? (Maximum 25 points)

3. How significant is the envisaged outcome in promoting priority issues in areas that are critical to Qatar, the regional and/or the world community? What may be the benefits of the proposed research activity to society? (Maximum 10 points)
4. How well conceived and organized is the overall proposed project? How sound is the experimental design? How appropriate are the statistical methods, if applicable, to analyze the data? How feasible is this project, given the resources and plan? (Maximum 10 points)
5. How sound is the evaluation plan? (Maximum 10 points)

<u>Research Team Qualification (25 points total)</u>

1. How well qualified are the principal investigators and the team to conduct the research (if appropriate, consider prior work)? (maximum 15 points)
2. Are the proposed staffing and facilities, including equipment and material available or requested, adequate to successfully execute the project? (maximum 10 points)

Bonus Points: Projects will receive 2 bonus points for collaboration with an institution outside of Qatar. Projects will receive 3 bonus points for intra-campus collaboration in Qatar or collaboration between an academic and a non-academic institution in Qatar. Proposals can receive a maximum of 3 bonus points total; a proposal cannot receive bonus points for both international collaboration and collaboration in Qatar.

Peer reviewers will provide narrative comments (not to exceed two pages) and the scores described above to QNRF. QNRF will make the reviews available to Principal Investigators, without identification of the peer reviewer. QNRF will then rank proposals within their disciplines, based on the technical review scores.

Step 4: Funding Allocation By Discipline and Project Funding

QNRF will provide a funding committee with summary statistics of the reviewed proposals. For example, the summary may include how many proposals were reviewed in each discipline, the distribution of scores among proposals, and the total amount requested in each category. The funding committee will then make recommendations to Qatar Foundation on how much of the total available grant budget to allocate to each discipline, but the funding committee will not review individual proposals or see the names/institutions of individual proposals.

QNRF will then fund proposals in rank order based on peer review scores by discipline, to the extent of available funding above a cut-off line.

At the discretion of QNRF Program Managers, site visits may be part of the review process. Unsuccessful applicants may apply again in the next cycle of the Program, taking into account the comments of the peer reviewers and QNRF on the relevance of the proposal to the national needs of Qatar.

11. Project Management and Research Products

11.1 Modifications to Terms of a Grant
During a project, the terms of the grant may be modified in ways that support advancement of the project. Grantees have discretion to re-budget within the cost of the grant. Unless otherwise stated in the grant or as noted below, the grantee is authorized to transfer funds among various budget lines within a category for allowable expenditures without prior QNRF approval. Prior written authorization from the QNRF is required for the following substantial changes:

- Transfer of the project effort to another institution
- Transfer of a PI in Qatar to an institution outside of Qatar
- Transfer of funds between institutions involved in the project
- Transfer of funds from the compensation of research staff category to other categories
- Extension of the grant agreement term with no additional funding
- Substantial change in objectives or scope
- Change in Principal Investigator, co-Principal Investigator, or other project leadership

Less than 25% of a grant may be carried into a subsequent year, if funding is left over at the end of a funding year. Above 25% requires a letter of justification and approval from QNRF.

11.2 Research Documents and Products
The following research products are required.

Interim Status Report. A status report of not more than 5 pages must be submitted to QNRF every six months, as well as at the end of the project. It should highlight expenditures, progress made, results obtained, problems or difficulties encountered, and deviations from the original objectives or research plan, if any. The reports will list all staff, with their positions, ranks, and institutional affiliations. They will also include financial overviews, detailing outlays by budget category and describing any significant deviations from anticipated expenditures (see below for related pre-authorization requirements). Additional payments of the grant (see the payment schedule) will be tied to the submission and acceptance of the progress reports. Grant payments will be delayed if the grantee fails to submit satisfactory and timely reports or if submitted reports indicate a substantial unexpended balance of grant funds on hand.

Yearly Report. Within 90 days after the end of a grant year, the Principal Investigator is required to submit a Yearly progress report. The Yearly Report should provide an overview of the achievements of the project, results obtained, problems or difficulties encountered, and deviations from the original objectives or research plan. It should also describe anticipated impact of the research results, and describe desired next steps. It will include information on all project findings and accomplishments, presented in terms accessible to a well-informed layman. Accomplishments include both work products, such as publications, reports, presentations, research methods, databases, patents, and licensing agreements, and outcomes, such as reductions in production costs or improvements in healthcare delivery.

Research Products. An electronic copy and two hard copies of each publication, report, or other documented product resulting from the grant should be sent to the QNRF Program Officer for archival purposes. Research products must be in accordance with QNRF's Intellectual Property Policy; this will be available online at a later date. Any publication, report, or other documented product resulting from a QNRF grant will include the following statement:

> This [publication, report, etc.] was made possible by a grant from the Qatar National Research Fund. The statements made herein are solely the responsibility of the author[s].

Abstracts. An abstract suitable for publication in QNRF's biannual survey of research in Qatar should be provided.

Participation in QNRF Seminars. PIs must be willing to participate in semi-annual or annual conferences or workshops organized by QNRF in order to share findings, lessons learned, and technical or policy barriers. QNRF may request presentations of progress made if needed at any time.

Post-Completion Reports. Grantees should also provide post-completion reports, if applicable, after the term of the grant, outlining any subsequent grant-derived accomplishments. Given ordinary lags in developing work products, e.g., publishing research findings or acquiring patents, and in establishing outcomes, it may be difficult to assess the full impact of funding without post-completion reporting.

11.3 Communication of Research Results

The QNRF will create and maintain a publicly accessible project database with links, as appropriate, to work products. For example, a web-based database could include the following: the program or project title, the name of the Principal Investigator, the start and end dates, a summary of findings, as approved by the Principal Investigator or project leader, a list of work products, and links to final reports, publications, and other documented work products. Principal Investigators must cooperate in providing this information.

11.4 Intellectual Property Policy

QNRF will follow the international model for intellectual property rights and revenue sharing for research outcomes and technology transfer. Please see the QNRF website www.qnrf.org for the Intellectual Property Policy.

12. Timeline - 1st Annual National Priorities Research Program Grant Cycle

The following dates represent the major milestones of the National Priorities Research Program:

April 23, 2007	Launch and public availability of National Priorities Research Program Request for Proposals.
Mid-May, 2007	Public Presentation and Q&A Sessions; QNRF will post Q&As from the sessions on its website.
May 1 – June 22, 2007	QNRF will respond to questions and inquiries from applicants submitted online, and post responses on its website. Applicants are advised to monitor the QNRF website for any final updates. QNRF will not accept phone calls or meetings to discuss the RFP with individual candidates, so that all candidates have the same information as posted online.
June 1, 2007	Detailed proposal submission instructions will be available online at www.qnrf.org.
June 23 – November, 2007	QNRF will not answer applicant questions or communicate with applicants during this time. All communication must be during the above period.
July 1, 2007	Letter of Intent Due
August 1, 2007	Proposals Due
November, 2007	Awards Announced and Grant Contracts Signed

Appendix 1
Examples of Focused Research Areas of Priority to Qatar

As part of the National Research Program, the QNRF seeks to support a broad variety of projects. These may include basic and applied research and projects in the arts, sciences, healthcare, engineering, and humanities. Proposals are invited in all of these disciplines.

Below is a list of examples of broad research areas relevant to Qatar's national needs for this first cycle of the NPRP. This list is largely based on a study by the Angle Technology Group of research areas relevant to Qatar.

The list is included for the purpose of sharing research ideas with potential researchers. Proposals for other research topics that the applicant believes are of importance to Qatar and can lead to 1) accumulation of valuable human capital or a sustainable research capability, or 2) answers to research questions of vital interest to Qatar, or 3) recognition of Qatar due to the regional or global significance of the proposed research are highly encouraged.

QNRF has developed more detailed lists that include examples of research themes and topics within certain research areas, namely: Industry and Environment, Computer Science and Information Technology, and Health and Biosciences after consultation with relevant stakeholders in Qatar. Applicants are welcome to visit these lists if they are interested. (www.qnrf.org)

INDUSTRY & ENGINEERING

- Energy/Oil &Gas/Petrochemicals
  - LNG transportation and economics
  - Robotics in hostile environments
  - Enhanced oil recovery
  - Utilization of sulfur
  - Reservoir modeling
  - Gas-to-Liquids process and economics
  - Process heat utilization
  - Gas flaring
  - Alternative energy
  - Fuel cells
  - Energy efficiency
  - Electricity demand management
  - Resource sustainability
  - Fluid separation
  - Carbon dioxide capture from flue gas
  - Production of gas hydrate as a means of gas utilization
  - Sea transportation of compressed natural gas
  - Improve oil & gas extraction efficiency
  - Remediation (pollutants and byproducts)
  - Robotics in hostile environments
- Materials
  - Corrosion
  - Non-destructive testing
  - Catalysis
  - Advanced polymers
  - Recycling
  - Alloys for high salinity environment
  - New materials and nanotechnology
- Aviation Operations
  - Airport Efficiency
  - Logistics Management
  - Training
  - Failure analysis

ENVIRONMENT
Water/Air/Soil

- Water
  - Supply
  - Distribution
  - Quality
  - Security
  - Desalination
  - Reuse and Irrigation
  - Importation
  - Cooling water
  - Water produced from oil and gas operation
  - Waste water management, treatment, reuse
- Air
  - Quality
  - Reduce stack emission
  - Ozone formation and level
- Soil & Ecosystem
  - Coastal and terrestrial ecosystems
  - High salinity marine ecology
  - Geospatial systems
  - Health effects of environmental pollutants
  - Biodiversity
  - Desertification
  - Impact of land reclamation

COMPUTER SCIENCE & INFORMATION TECHNOLOGY

- Computing on a national level
- Oil and gas data mining
- Integrated GIS graphic information systems
- Cyber security
- Robotics
- Networks
- Language translation
- Wireless networking
- Media
- National Digital Roadmap

HEALTH & LIFE SCIENCES

- Diabetes
- Obesity
- Cardiovascular diseases
- Women's and children's health
- Genomic factors on health of individuals and population
- Cancer (breast, colon, and hematological malignancies)
- Health and Environmental Pollution

EDUCATION

- E-learning
- Open systems
- Distance learning
- Virtual classrooms
- Software
- Methods and Systems

PUBLIC POLICY & MANAGEMENT

- Economic diversification
- Econometrics
- Intellectual property protection
- Risk Management and industrial safety
- Growth management
- Urban and regional planning, sustainable cities, population dynamics
- Media
- Aviation operations
- Technology innovation and commercialization
- Advancement of Women
- Operations Research
- Enterprise risk management
- Leadership, organizational behavior, communication
- Innovation and entrepreneurship
- Management in Qatar

ARTS & HUMANITIES

- Design
- Shariah
- Ethics of new technologies

SOCIAL SCIENCES

- Regional history, culture and law
- Regional political science
- East/West understanding
- Behavior modification
- Diet and exercise
- Motivation
- Conflict resolution

Appendix 2
Code of Practice

1. Code of Practice for Consideration of Research Proposals

a. This Code of Practice describes the standards of transparency by which the Qatar National Research Fund (QNRF) abides in administering applications for research awards and other proposals for support, and embodies the principles of equity, integrity and confidentiality for all who are involved in the assessment of proposals. The Code is also intended to act as guidance to peer reviewers in discharging the responsibilities placed on them in assessing proposals, and sets out the proper conduct expected of them.

2. Information for applicants

a. **Application procedures** For each of its schemes for research awards, QNRF issues guidelines on the information to be supplied by applicants in support of bids for funds, details of the criteria against which the application will be assessed, and the process and timescale for assessment of the application. Any other details of the program may be clarified in publicly posted Questions and Answers on the QNRF website.

b. **Data Protection** Applicants are required to sign the application to indicate that the information provided therein is, to the best of their knowledge, complete and accurate. Applicants should be aware that information they provide would be stored and circulated as necessary for the assessment procedures to be followed. Successful applicants should be aware that the information they provide on the application form may be copied to the relevant authorized officer in their employing institution as necessary for the award procedures to be followed, and information on the status of their award may be made available to the relevant authorized officer in their employing institution by QNRF as necessary for the conditions of award to be fulfilled. Application forms will be retained for ten years in the case of successful applications, and five years in the case of unsuccessful applications, and may be consulted by QNRF in the event of future applications being submitted. Details of award holders (including name, institution, project details and amount of award) will be used to compile published lists of award-holders that will be made available on the Internet, and to produce statistical and historical information on QNRF awards. Signing the application form constitutes the applicant's agreement to all terms, conditions, and notices contained in the Notes for Applicants. Once QNRF has received the final applications, it will respect the integrity of the applications and not alter them in any way.

c. **Data monitoring** Personal information provided by applicants will be used for monitoring and statistical purposes only, and at no stage will it form any part of the assessment process.

d. **Equal opportunity** QNRF is committed to a policy of equal opportunity in that applicants will receive equal treatment, regardless of race, color, religion, gender, age, nationality (except where the conditions of the scheme specify otherwise) or disability.

e. **Ethics policy** QNRF requires the research it funds to be conducted in an ethical manner. The following considerations apply to all proposals:
   i. accurate reporting of findings and a commitment to enabling others to replicate results where possible;
   ii. fair dealing in respect of other researchers and their intellectual property;
   iii. proper employment conditions for research staff;
   iv. honesty to research staff and students about the purpose, methods and intended and possible use of the research and any risks involved;
   v. confidentiality of information supplied by research subjects and anonymity of respondents (unless otherwise agreed with research subjects and respondents);
   vi. independence and impartiality of researchers to the subject of the research.

Additionally, proposals may raise one or more of the following considerations: the involvement of human participants; the involvement of human remains (e.g., traceable to living descendants); the use of non-human animals; destructive analysis of historic artifacts; research that may result in damage to the natural or historic environment; and the use of sensitive social, economic or political data. Wherever necessary, appropriate consent should be obtained from or on behalf of participants or others affected by the research. Applicants should indicate whether their proposed research raises any special ethical issues, and whether the relevant authority has approved their application. Independent researchers without access to formal ethical scrutiny and approval should briefly describe any special ethical issues, and explain how they will be addressed.

f. **Assessment process** Appropriate experts judge all applications on their academic merit through a stringent process of peer review. Recommendations are passed to the relevant awarding committee for final decision on awards. QNRF will respect the integrity of the reviews and not alter them in any way.

g. **Outcome of applications** Applicants are informed by letter of the outcome of their application. Feedback, in the form of anonymous comments from external peer reviewers, will be provided for applicants. QNRF is regretfully unable to enter into correspondence concerning the decision of the awarding committee. Applicants are informed in the notes of guidance whether feedback can be expected as a feature of the scheme.

h. **Conditions of awards** Recipients of awards are made aware of the regulations governing the scheme in which they have been successful and are required to adhere to those regulations.

i. **Appeals** The competition for research awards is intense and many high quality applications may not receive support. All applications receive careful scrutiny by the peer reviewers, in the context of competing claims on available funding. Appeals may therefore not be made against the academic judgment of the QNRF's peer reviewers, panels, or Committees. The sole ground on which an appeal may be made is one of improper procedure. Anyone wishing to make an appeal against a decision should write to the Director of QNRF no later than two months after the result of the competition is announced, citing the specific decision and setting out clearly the substantive basis of the appeal. Only applicants themselves may appeal, though they may include supporting letters as relevant. The Chief Executive will respond in writing within 30 days. There are two possible grounds for one further stage of appeal: either improper procedure in the investigation of the original appeal; or the availability of substantial relevant information which for good reason was not made known to the Director at the time of the investigation.

3. Information for Peer Reviewers

a. **Confidentiality** Those who undertake the assessment of applications are required to give assurance that all information which they acquire in the discharge of their duties be kept confidential and not be transmitted to any persons other than in accordance with the prescribed procedures for the selection process. All reasonable steps must be taken to ensure that such information is kept in a secure place and in due course disposed of in a secure fashion (or returned to QNRF). Information provided to peer reviewers in an application for funding may only be used for the purposes of evaluating the proposal in accordance with QNRF's guidelines.

b. **Conflict of interest** Those who undertake the responsibility of assessing applications for funds, either in writing or through membership of awards committees, are required to declare actual or potential conflicts of interest and observe the following guidelines:

   i. *References* Peer reviewers, including members of awards committees, shall not act as referees for individual candidates in any of QNRF's grant-giving schemes in which they are involved in any capacity.

   ii. *Institutional affiliation* Peer reviewers shall not participate in the evaluation of any proposal emanating from their own institution.

   iii. *Other connections* Where an application involves a former pupil, close colleague or co-researcher, a family member, or a person with whom there is or has been a current or prior relationship, peer reviewers are required to declare any conflict of interest to the relevant QNRF officer so that the proposal can be redirected (in the case of research and conference grants), and peer reviewers, including those involved in the assessment of research posts, shall abstain from participating in the evaluation of that particular proposal.

   iv. *Peer reviewers as applicants* Fellows who wish to apply for QNRF support during the period in which they are serving in any capacity as an peer reviewer must abstain from any involvement in the competition to which they are applying, that is, they may not assess or comment or vote on any application in that round of the competition.

v. If peer reviewers are unsure whether their ability to assess a proposal is compromised in any way, they should inform QNRF of the relevant circumstances so that guidance can be sought on individual cases. A log of such incidents shall be retained for the regular scrutiny of the QNRF's Steering Committee and/or Governing Board.

c. **Fair evaluation** Peer reviewers are normally drawn from subject-specific experts within the international academic community, and it is expected that they will be able to evaluate the proposals sent to them. In cases where individual peer reviewers feel unable to offer an informed view on a proposal, they may request that proposals be sent confidentially to other members of QNRF so that they may consult about the merits of the proposal. In all cases, peer reviewers must submit the request for additional assessment through the office to ensure that the chosen advisers receive the necessary instructions about assessment criteria particular to the relevant scheme, and a copy of this Code of Practice. Any supplementary advisers are required to abide by its provisions.

This Code of Practice is adapted for QNRF's use from The British Academy's Code of Practice. QNRF expresses gratitude to The British Academy for its permission to do so.

APPENDIX C

Overview of QNRF's Strategy Statements

Qatar Foundation's Mission

Qatar Foundation's mission is to prepare the people of Qatar and the region to meet the challenges of an ever-changing world, and to make Qatar a leader in innovative education and research. To achieve that mission, Qatar Foundation supports a network of centers and partnerships with elite institutions, all committed to the principle that a nation's greatest natural resource is its people. Education City, Qatar Foundation's flagship project, is envisioned as a Center of Excellence in Education and Research that will help transform Qatar into a knowledge-based society (Qatar Foundation, undated).

QNRF's Vision

The Qatar Foundation envisions research as a catalyst for expanding and diversifying the country's economy; enhancing the education of its citizens and the training of its workforce; and fostering improvements in the health, well-being, environment, and security of its own people and those of the region. In striving toward this vision, Qatar will distinguish itself within the region and world as a cosmopolitan nation that embraces scholarly excellence, innovation, creativity, inclusiveness, and merit (Greenfield et al., 2008).

QNRF's Mission

QNRF will advance knowledge and education by supporting original, competitively selected research in the physical, life, and social sciences; engineering and technology; the arts; and humanities. It will provide opportunities for researchers at all levels, from students to professionals, in the private, public, and academic sectors (Greenfield et al., 2008).

QNRF's Goals[1]

1. Build human capital.
2. Fund research in the interest of Qatar, the region, or the world.
3. Raise Qatar's international profile in research.

[1] Greenfield et al., 2008.

Guiding Principles for QNRF's Research Programs

- QNRF programs should be geared toward creating a research culture in Qatar, focusing on building human capital.
- Program designs should include attractive incentives for researchers and institutions.
- Programs need one set of policies that can accommodate research in different parts of the world.
- Programs require buy-in (support and feedback) from participating institutions in order to effectively meet those institutions' needs.
- QNRF should learn from its own experiences in designing programs and should use that knowledge to make improvements.
- QNRF policies should be clear, transparent, and consistently applied.

National Priorities Research Program Objectives

- Attract, develop, and retain faculty at Qatar University and Education City and researchers at other institutions in Qatar.
- Provide incentives to build an institutional infrastructure in Qatar that supports research.
- Advance knowledge and technology transfer within Qatar through collaboration with institutions outside of Qatar.

References

Alfred P. Sloan Foundation, "Sloan Research Fellowships," website, 2008. As of June 20, 2012: http://www.sloan.org/fellowships

Anderson, Philip, "Complexity Theory and Organization Science," *Organization Science*, Vol. 10, No. 4, May–June 1999, pp. 216–232.

Buchanan, Phil, *Foundation Governance: The CEO Viewpoint,* Cambridge, Mass.: The Center for Effective Philanthropy, February 2004. As of June 20, 2012:
http://www.effectivephilanthropy.org/assets/pdfs/CEP_FoundationGovernance.pdf

Burns, Lawron R., and Douglas R. Wholey, "Adoption and Abandonment of Matrix Management Programs: Effects of Organizational Characteristics and Interorganizational Networks," *The Academy of Management Journal*, Vol. 36, No. 1, February, 1993, pp. 106–138.

Cecchine, Gary, Richard E. Darilek, Margaret C. Harrell, Michael Mattock, Shelly Culbertson, David S. Ortiz, Nicholas W. Maynard, S. R. Bohandy, Edward Balkovich, and Bruce Nardulli, *Sustaining the Qatar National Research Fund*, Santa Monica, Calif.: RAND Corporation, TR-1125-QF, 2012.

The Center for Effective Philanthropy, *Toward a Common Language: Listening to Foundation CEOs and Other Experts Talk About Performance Measurement in Philanthropy*, Boston, Mass.: The Center for Effective Philanthropy, February 2002. As of April 4, 2011:
http://www.effectivephilanthropy.org/assets/pdfs/CEP_TowardsACommonLanguage.pdf

———, *Indicators of Effectiveness: Understanding and Improving Foundation Performance: Report on the Foundation Performance Metrics Pilot Study*, Boston, Mass.: The Center for Effective Philanthropy, August 2002. As of April 4, 2011:
http://www.effectivephilanthropy.org/assets/pdfs/CEP_IndicatorsOfEffectiveness.pdf

DARPA, "About," website, undated a. As of June 20, 2012:
http://www.darpa.mil/About.aspx

———, "Contract Management," website, undated b. As of July 18, 2012:
http://www.darpa.mil/Opportunities/Contract_Management-Contract_Management.aspx

———, "Creating and Preventing Strategic Surprise," website, undated c. As of April 4, 2011:
http://www.darpa.mil/

Ford, Robert C., and W. Alan Randolph, "Cross-Functional Structures: A Review and Integration of Matrix Organization and Project Management," *Journal of Management*, Vol. 18, No. 2, June 1992, pp. 267–277.

Greenfield, Victoria, et al., *Design of the Qatar National Research Fund: An Overview of the Study Approach and Key Recommendations*, Santa Monica, Calif.: RAND Corporation,TR-209-QF, 2008. As of April 4, 2011:
http://www.rand.org/pubs/technical_reports/TR209.html

Johnson, Valen E., "Statistical Analysis of the National Institutes of Health Peer Review System," *Proceedings of the National Academy of Sciences*, Vol. 105, No. 32, August 12, 2008, pp. 11076–11080.

Lopatto, D., "Undergraduate Research Experiences Support Science Career Decisions and Active Learning," *CBE Life Science Education*, Vol. 6, No. 4, Winter 2007, pp. 297–306.

Mowery, David C., Richard R. Nelson, Bhaven N. Sampat, and Arvids A. Ziedonis, "The Growth of Patenting and Licensing by U.S. Universities: An Assessment of the Effects of the Bayh-Dole Act of 1980," *Research Policy,* Vol. 30, No. 1, 2001, pp. 99–120.

Nagda, Biren A., Sandra R. Gergerman, John Jonides, William von Hippeo, and Jennifer S. Lerner, "Undergraduate Student-Faculty Research Partnerships Affect Student Retention," *The Review of Higher Education*, Vol. 22, No. 1, 1998, pp. 55–72.

National Institutes of Health, "About NIH," website, last reviewed December 5, 2011. As of June 20, 2012: http://www.nih.gov/about

National Science Foundation, "National Science Foundation History," undated a. As of June 20, 2012: http://www.nsf.gov/about/history/

———, "Research Experiences for Undergraduates (REU)," undated b. As of May 24, 2011: http://www.nsf.gov/funding/pgm_summ.jsp?pims_id=5517&from=fund

———, "Research Experiences for Undergraduates (REU): Supplements and Sites: Program Solicitation," NSF 03-577, undated c. As of April 4, 2011: http://www.nsf.gov/pubs/2003/nsf03577/nsf03577.pdf

———, "National Science Foundation Investing in America's Future Strategic Plan FY 2006–2011," NSF 06-48, September 2006. As of April 4, 2011: http://www.nsf.gov/pubs/2006/nsf0648/nsf0648.jsp

———, "Research Experiences for Undergraduates (REU): Sites and Supplements," NSF 07-569, last updated November 7, 2006. As of April 4, 2011: http://www.nsf.gov/pubs/2007/nsf07569/nsf07569.htm

Qatar Academy, "Board of Governors," website, undated. As of April 4, 2011: http://www.qa.edu.qa/about-qatar-academy/board-of-governors

Qatar Foundation, "Vision and Mission," website, undated. As of June 20, 2012: http://www.qf.org.qa/clubhouse/output/page77.asp

Qatar Science and Technology Park, "FAQ," website, undated. As of June 20, 2012: http://www.qstp.org.qa/home/about-us/faq

QSTP—*See* Qatar Science and Technology Park.

Sidra Medical and Research Center, "Governance," website, undated. As of May 12, 2012: http://www.sidra.org/en/about/governance/Pages/default.aspx

University of Pittsburgh, University Honors College, "Brackenridge Summer Research Fellowships," website, 2012. As of May 12, 2012: http://www.honorscollege.pitt.edu/node/326

U.S. Department of Health & Human Services, Office of Extramural Research, National Institutes of Health, website, updated February 24, 2012. As of June 20, 2012: http://grants.nih.gov/grants/oer.htm